Immuno-Oncology

Progress in Tumor Research

Vol. 42

Immuno-Oncology

Volume Editors

Olivier Michielin Lausanne
George Coukos Lausanne

8 figures, 5 in color, and 11 tables, 2015

KARGER Basel · Freiburg · Paris · London · New York · Chennai · New Delhi · Bangkok · Beijing · Shanghai · Tokyo · Kuala Lumpur · Singapore · Sydney

Progress in Tumor Research
Formerly published as 'Progress in Experimental Tumor Research'

Olivier Michielin, MD, PhD
Associate Professor
Department of Oncology
University Hospital of Lausanne (CHUV)
Ludwig Cancer Research Center
University of Lausanne
Swiss Institute of Bioinformatics, Lausanne
Rue du Bugnon 46 – BH09-733
CASE POSTALE 5
1011 Lausanne (Switzerland)

George Coukos, MD, PhD
Professor
Director, Department of Oncology
University Hospital of Lausanne (CHUV)
Director, Ludwig Cancer Research Center
University of Lausanne
Rue du Bugnon 46 – BH09-701
CASE POSTALE 5
1011 Lausanne (Switzerland)

Library of Congress Cataloging-in-Publication Data

Immuno-oncology / volume editors, Olivier Michielin, George Coukos.
p. ; cm. -- (Progress in tumor research, ISSN 2296-1895 ; vol. 42)
Includes bibliographical references and indexes.
ISBN 978-3-318-05589-4 (hard cover : alk. paper) -- ISBN 978-3-318-05590-0 (electronic version)
I. Michielin, Olivier, editor. II. Coukos, George, editor. III. Series: Progress in tumor research ; v. 42. 2296-1895
[DNLM: 1. Immunotherapy--methods. 2. Neoplasms--therapy. W1 PR668T v.42 2015 / QZ 266]
RM370
615.3'7--dc23

2015028888

Bibliographic Indices. This publication is listed in bibliographic services, including Current Contents®.

www.karger.com
Printed in Germany on acid-free and non-aging paper (ISO 9706) by Kraft Druck, Ettlingen
ISSN 2296–1895
e-ISSN 2296–1887
ISBN 978–3–318–05589–4
e-ISBN 978–3–318–05590–0

KARGER 125 YEARS

Contents

Preface

Over the last decade, immuno-oncology has witnessed an astonishing pace of discovery and innovation translating into unprecedented successes in the clinical setting, arguably representing one of the most profound and transforming revolution in the history of cancer therapy. A key aspect of this growing success is the path that led us there. Rather than empirical, rationally designed and molecularly defined clinical interventions from sound basic, translational and clinical research, paved the way up to drug registration. This rational approach holds the promise of further new exciting breakthroughs originating from ongoing research and discoveries that have not yet reached the clinic.

In this book, we will provide a concise, yet very accurate outline of the main developments in major tumor types including melanoma, lung, breast, brain and renal cell cancers. In addition, transversal Chapters that describe the commonalities of some of the therapeutic strategies are provided to cover topics like immune checkpoint biology, T cell engineering or rational combination therapies. Each Chapter has been authored by senior key opinion leaders in their respective fields to provide the most up-to-date view on cancer immuno-oncology. To reflect on the key translational aspect of immuno-oncology, all chapters are making explicit connections between basic science discoveries and the resulting translational therapeutic strategies.

We believe this book will be an invaluable source of information for scientists interested into the translation of basic immunology into the clinical practice, as well as for clinician interested in deepening their knowledge of current and upcoming immune strategies in the fight against cancers.

We hope the reader will share our enthusiasm for this field of research and certainly wish him an enjoyable reading!

Olivier Michielin, Lausanne
George Coukos, Lausanne

Michielin O, Coukos G (eds): Immuno-Oncology.
Prog Tumor Res. Basel, Karger, 2015, vol 42, pp 1–10 (DOI: 10.1159/000436988)

Immunotherapies in Early and Advanced Renal Cell Cancer

Benjamin Kasenda · James Larkin · Martin Gore
Medical Oncology, Royal Marsden Hospital, London, UK

Abstract
The development of new immunomodulatory monoclonal antibodies targeting the CTLA-4 or PD-1 axis has led to a revival of research on immunotherapies in solid tumours including renal cell cancer (RCC). The initial results observed with these monoclonal antibodies in the treatment of advanced melanoma have resulted in considerable interest in this treatment strategy in all tumour types. Preliminary data of these new antibodies in advanced RCC are promising and they have good safety profiles. Response rates are low but durable tumour control has been observed in some patients. However, at the moment there is no evidence that targeting the CTLA-4 or PD-1 axis provides a substantial clinical benefit compared to established treatment with tyrosine kinase or mTOR inhibitors. There are also no reliable predictive markers. At the moment, several randomised trials have been initiated to investigate the new immunomodulatory antibodies either as single agents or in combination with anti-VEGF targeted therapy. Vaccines have continued to be investigated in advanced and adjuvant settings. No trial has so far established vaccines as a standard treatment in either situation. There are still large randomised trials ongoing investigating the potential benefit of a vaccine in combination with standard tyrosine kinase inhibitor therapy. In this chapter we will summarise selected studies on immunotherapy in advanced RCC with a focus on anti-PD-1, anti-PD-L1, and anti-CTLA-4 antibodies. We will also touch briefly on the adjuvant situation and tumour vaccines.

Vascular endothelial growth factor (VEGF) inhibitors and mTOR inhibitors are now considered standard treatment in advanced or metastatic renal cell cancer (RCC) [1]. Previously, in Europe immunotherapy with interferon-alpha (IFN-α) was the standard treatment for advanced renal cancer, and in the USA interleukin-2 (IL-2) was used in fit patients. Treatment with IFN-α was based on several randomised trials [2–4] and a Cochrane meta-analysis from 2006 which confirmed a statistically significant survival benefit for IFN-α (OR for death at 1 year = 0.56, 95% CI 0.40–0.77) compared to non-immunotherapies [5]. The absolute survival benefit was about 15% at 1 year with an increase in median survival benefit in the region of 4 months. Studies of IL-2 suggested that treatment with high-dose (HD) intravenous schedules could lead to durable remission rates in 5–10% of patients [6–9], but there was no advantage based on randomised comparisons to controls not receiving HD IL-2. A randomised trial was conducted to test whether the combination of IFN-α, low-dose IL-2 and 5-fluorouracil is superior to IFN-α alone. The response rate in the combination arm was higher (23 vs. 16%), but there was no difference in progression-free (PFS) or overall survival (OS) [10]. Currently, particularly in the USA, some units still use HD IL-2 in selected patients, and IFN-α treatment is still part of European guidelines, but only in combination with bevacizumab based on two randomised trials [1, 11, 12]. The question remains whether the combination of bevacizumab with IFN-α is superior to bevacizumab alone.

New immunomodulatory monoclonal antibodies that target the inhibitory cross-talk between tumour cells and the immune system via the CTLA-4 or PD-1 axis have recently become available. These new antibodies have revolutionised the treatment of metastatic melanoma [13–16] and have also shown promising activity in heavily pre-treated Hodgkin lymphomas [17] and squamous cell lung cancer [18]. In this chapter we will summarise selected studies on immunotherapy in advanced RCC with a focus on anti-PD-1, anti-PD-L1 and anti-CTLA-4 antibodies. We will only briefly discuss adjuvant therapy and tumour vaccines.

Anti-CTLA-4 Therapy

Two CTLA-4 antibodies, namely ipilimumab [19] and tremelimumab [20], have been investigated in non-randomised studies enrolling 51 and 19 patients with RCC, respectively. Yang et al. [19] treated 2 cohorts with ipilimumab: cohort A had a loading dose of 3 mg/kg followed by 1 mg/kg (21 patients) every 3 weeks, and cohort B received all doses at 3mg/kg every 3 weeks (40 patients). Treatment was given for up to 1 year until disease progression or significant toxicities. Major toxicities were enteritis and endocrine deficiencies, with 33% experiencing grade 3/4 immune-mediated side effects. Overall, 6 out of 51 (12%) patients achieved a partial remission (1 in cohort A, 5 in cohort B) with response durations ranging from 7 to 21 months. In the smaller phase I study, Rini et al. [20] treated patients with a combination of tremelimumab (6, 10 or 15 mg/kg) plus sunitinib (50 mg, 4 weeks on 2 weeks off treatment, or 37.5 mg daily without interruptions). Among patients who received continuous sunitinib 37.5 mg daily, 1 of the 7 patients who received tremelimumab 10 mg/kg plus sunitinib suffered a sudden death. Overall, rapid-onset renal failure was the most common dose-limiting toxicity (10%). Nine out of 21 patients (43%) who were evaluable for response achieved partial responses. This combination was not recommended for further development because of the unexpected serious renal toxicity [20].

Anti PD-1 and PD-L1 Therapy

Topalian et al. [21] reported a phase I trial of nivolumab (PD-1 antibody) in various solid tumours (n = 296) including RCC. Overall, in 14% of patients, grade 3/4 drug-related adverse events occurred, including three deaths from pulmonary toxicity. Thirty-three patients with RCC were treated, 17 with 1 mg/kg and 16 with 10 mg/kg; 6% discontinued treatment due to adverse events. The respective response rates in the different dosing groups were 24% and 31%, with the same proportion of patients achieving stable disease (24% and 31%, respectively) [21]. In a recent update reporting a longer follow-up, the median PFS was 7.3 months and median OS was not reached. Even after discontinuation of nivolumab, durable responses were seen [22]. In a randomised phase II trial including patients pre-treated with anti-VEGF drugs (n = 168), three doses of nivolumab were compared: 0.3, 3 and 10 mg/kg once every 3 weeks [23]. The median PFS was 2.7, 4.0 and 4.2 months, respectively; response rates were about 20% in each group. The respective median OS in the 3- and 10-mg group was 25.5 and 24.7 months, respectively, and 18.2 months in the group receiving 0.3 mg. A durable response continuing for >24 months occurred in 14 out of 35 patients who achieved a response. The most common treatment-related adverse event was fatigue (24–35%) and, overall, 19 patients experienced grade 3/4 treatment-related toxicities [23]. The authors concluded that nivolumab demonstrated anti-tumour activity with the potential of long-term tumour control and toxicities were manageable across the three doses studied. No dose-response relationship was detected as measured by PFS [23]. A phase I trial of the PD-L1 antibody MPDL3280A in 53 patients with RCC treated 2 patients with 3 mg/kg, 12 patients with 10 mg/kg, 18 patients with 15 mg/kg and 21 patients with 20 mg/kg. Treatment was administered every 3 weeks for up to 1 year [24]. Overall, grade 3/4 adverse events occurred in 43% of patients with hypophosphatemia, fatigue, dyspnoea and hyperglycaemia being the most common side effects. However, only 13% of these events were considered attributable to the study drug and, of note, no grade 3/4 pneumonitis or diarrhoea occurred. One case of myasthenia gravis was reported and, retrospectively, this patient was found to have anti-acetylcholine receptor antibodies prior to starting therapy. The 24-week PFS was 50% and there was some evidence for a positive correlation between the tumour PD-L1 status and clinical efficacy [24]. This PD-L1 antibody (MPDL3280A) was further investigated in combination with bevacizumab in another phase I trial which included patients with various solid tumours (n = 71), 12 of whom had RCC [24]. Ten of these patients were receiving first-line treatment. Patients with RCC received MPDL3280A 20 mg/

kg in combination with bevacizumab 15 mg/kg every 3 weeks. An objective response was observed in 30% of the RCC patients. Overall, grade 3/4 toxicities occurred in 49% of patients, with abdominal pain, pneumonia and hypertension being the most frequent [25]. Brahmer et al. [26] reported a phase I trial investigating another PD-L1 antibody (BMS-936559) in patients with various solid tumours (n = 207), including 17 patients with advanced RCC. Overall, treatment was well tolerated, with grade 3/4 toxicities only occurring in 9% of all patients. All patients with RCC were treated with 10 mg/kg every 2 weeks. An objective response was observed in 2 patients (12%), with respective response durations of 4 and 17 months; 53% of patients remained progression-free at 24 weeks. Seven of 17 (41%) renal cancer patients showed stabilization of the disease for at least 24 weeks [26].

In summary, ipilimumab, anti-PD-1 and anti-PD-L1 antibodies have shown promising clinical activity and, although response rates to single-agent therapy are not high (20–25%), there seems to be a good chance for long-term tumour control in some patients. The safety profile is similar to that previously encountered with these agents. However, the onset of acute renal failure with the combination of tremelimumab and sunitinib was not expected. Some studies have suggested a positive correlation between tumour PD-L1 expression and response to anti-PD-1 or anti-PD-L1 therapy. Interestingly, in pre-specified subgroup analyses from a randomised phase III trial in advanced melanoma (nivolumab vs. dacarbazine), patients with a positive PD-L1 status seemed to have a higher response compared to PD-L1 negatives (53% vs. 33%); however, OS was very similar [16]. No interaction analyses were conducted; therefore, whether PD-L1 expression is a reliable predictive marker in melanoma remains to be determined.

In RCC, there is some evidence that PD-L1 expression may have some prognostic impact on survival; increased PD-L1 expression retrieved from archival tissue was inversely correlated with survival [27–29]. However, whether PD-L1 expression on tumour tissue in RCC could be a reliable predictive biomarker is subject to further research, especially regarding the standardization of expression level analyses and definition of appropriate cut-offs.

Combinations

Anti-CTLA-4 and Anti-PD-1

Hammers et al. [30] presented a phase I trial investigating the combination of nivolumab and ipilimumab in patients with RCC. This trial also includes a cohort of patients treated with a combination of nivolumab plus sunitinib or pazopanib (see below) [31]. Previously treated or treatment-naïve patients were randomised to receive either 4 cycles of nivolumab 3 mg/kg plus ipilimumab 1 mg/kg (N3 + I1) or nivolumab 1 mg/kg plus ipilimumab 3mg/kg (N3 + I1) every 3 weeks; both groups received maintenance nivolumab 3 mg/kg every 2 weeks until progression or intolerable toxicity [30]. There were 21 patients in the first group and 23 in the second. Treatment-related adverse events were seen in 39 patients (89%), 8 (18%) discontinued treatment due to adverse events, 6 of whom were treated with N1 + I3. Grade 3/4 adverse events occurred in 20 patients (46%), most of these (14 patients) were treated with N1 + I3. Response rates were 43% (N3 + I1) and 48% (N1 + I3) with a median duration of response of 7.2 months in the N3 + I1 group; the median was not reached in the N1 + I3 group. The median PFS was very similar for both groups: 8.4 months (N3 + I1) versus 8.8 months (N1 + I3) [30]. These encouraging results led to the initiation of a randomised phase III trial to test the superiority of the N3 + I1 combination compared to sunitinib in treatment-naïve patients (CheckMate 214, NCT02231749).

Anti PD-1 with Tyrosine Kinase Inhibitors

Amin et al. [31] reported a phase I trial of a combination of nivolumab (2 mg/kg initial and 5 mg/kg expansion cohort; 3 times weekly) with either sunitinib (50 mg, 4 weeks on and 2 weeks off treatment) or pazopanib (800 mg daily, continuously). Thirty-four patients with RCC were enrolled, 14 received nivolumab in combination with sunitinib (7 in the expansion cohort) and 20 received nivolumab 2 mg/kg plus pazopanib. There were 4 dose-limiting toxicities in the pazopanib arm, which was therefore closed and no expansion cohort was started; these toxicities were elevated ALT/AST (3 patients) and fatigue (1 patient). Overall, grade 3/4 adverse events were observed in 73% of patients in the sunitinib arms and in 60% in the pazopanib arm. The most common grade 3/4 adverse events in the sunitinib arms included elevated ALT (18%), hypertension (15%) and hyponatraemia (15%). In the pazopanib arm, elevated ALT/AST (20% each) and fatigue (15%) were the most frequent grade 3/4 adverse events [31]. One patient treated with nivolumab 5 mg/kg plus sunitinib developed grade 3 pneumonitis. In those patients treated with the sunitinib combination, 8 patients (24%) had to discontinue treatment because of grade 3/4 toxicities (1 treated with nivolumab 1 mg/kg, and 7 with nivolumab 5 mg/kg). Of the patients treated with the pazopanib combination, 20% had to discontinue because of toxicities. The overall response rate was similar, 52% for sunitinib and 45% for pazopanib patients with the duration of response ranging from 2.8 to 12.4 months (sunitinib) and 2.8 to 15.9 months (pazopanib). Overall, the PFS at 24 weeks was 78% for patients treated in the sunitinib arm and 55% for those in the pazopanib arm [31].

In summary, the combination of nivolumab (3 mg/kg) and ipilimumab (1 mg/kg) seems to be promising and is currently being investigated in a phase III trial that compares this combination to standard first-line sunitinib treatment. Based on one phase I study, sunitinib seems be the preferred partner to be combined with nivolumab. However, even with this drug about a quarter of patients had to stop therapy because of toxicities, which may question the feasibility of the combination of tyrosine kinase inhibitors and anti-PD-1 antibodies.

Other Immunotherapies for Advanced Disease

Oudard et al. [32] reported a phase II study of TG4010 (a vaccine based on a modified vaccinia virus expressing MUC1 and IL2) in combination with cytokines as first-line therapy in 37 patients with metastatic RCC. No objective clinical responses were observed. Five of the 27 evaluable patients (18%) had stable disease for >6 months with TG4010 alone, and 6 of 20 patients (30%) had stable disease for >6 months with TG4010 plus cytokines [32]. Another phase III trial tested whether the modified vaccinia Ankara encoding the tumour antigen 5T4 (MVA-5T4) could improve OS if added to standard first-line therapy. This double-blind trial recruited 733 patients but no survival impact was observed. A subgroup analysis suggested that patients with a higher 5T4-specific antibody response benefited from the experimental treatment, but further development of MVA-5T4 was not pursued [33]. One ongoing phase III trial is investigating the addition of autologous dendritic cells to sunitinib; the first results may be available in spring 2016 (NCT01582672).

Two trials are currently investigating the effect of tumour-associated peptides in advanced RCC. The compound IMA901 is a synthetic 'off-the-shelf' vaccine consisting of 10 different tumour-associated peptides. In a recent phase II trial, patients received IMA901 and granulocyte-macrophage colony-stimulating factor with or without cyclophosphamide (NCT00523159); the results of this trial are awaited. A phase III trial has investigated the addition of IMA901 to sunitinib for first-line treatment in advanced/metastatic RCC. The first results are awaited in the second half of 2015 (NCT01265901).

In summary, immunotherapies based on vaccines have not shown a significant clinical benefit in the advanced setting so far. Results from ongoing trials are awaited, especially those in combination with standard sunitinib treatment.

Adjuvant Studies

Vaccines

Several randomised trials have been conducted to test whether adjuvant immunotherapy-based treatment can prolong survival after resection of the primary RCC. Two small trials investigated autologous irradiated tumour cells in combination with Bacillus Calmette-Guérin, but none showed a clinical benefit [34, 35]. In a large randomised phase III trial, 558 patients scheduled for radical nephrectomy were enrolled at 55 centres in Germany and randomised to receive an autologous tumour cell vaccine or observation [36]. Randomization was performed before surgery and finally 379 patients (177 in the vaccination group, 202 in the control group) with a tumour ≥2.5 cm were included in the intention-to-treat analysis. PFS was significantly prolonged in the vaccination group (hazard ratio: 1.58 in favour of the vaccination group,

95% CI 1.05–2.37), but OS has not been reported as being prolonged [36]. Another randomised phase III trial investigated vitespen, also an autologous vaccine; the vaccine was compared to observation alone. Eight hundred and eighteen patients were enrolled and randomised from 118 centres in North America and Europe. There was no difference regarding the primary endpoint recurrence-free survival or OS [37].

IFN-α-Based Therapies
Encouraged by the activity of IFN-α in advanced or metastatic RCC, 2 large randomised trials investigated adjuvant treatment with IFN-α compared to observation alone. However, neither of them showed any clinical benefit [38, 39]. High-dose IL-2 similarly has not been shown to have any meaningful benefit in the adjuvant setting [40]. Combining low doses of IL-2 and IFN-α was also of no benefit, although a post hoc subgroup analysis suggested some benefit for younger patients with small tumours [41]. The combination of IFN-α, IL-2 and fluorouracil did not show any benefit in a recently published EORTC trial [42]; it has even been suggested that this combination might have a detrimental effect on OS as compared to observation only [2].

In summary, there is no convincing evidence for any adjuvant immunotherapy having a beneficial effect following radical nephrectomy for RCC.

Conclusion

The development of new immunomodulatory antibodies targeting the CTLA-4 or PD-1 axis has led to a revival of research on immunotherapies in solid tumours, including RCC. Research on anti-tumour vaccines has continued in parallel. The preliminary data on these new antibodies are promising with good safety profiles and durable disease control, even though the overall response rates are low. Currently, there is no evidence that targeting the CTLA-4 or PD-1 axis provides a substantial clinical benefit compared to established treatment with tyrosine kinase or mTOR inhibitors. However, this may change with key trials ongoing, which are investigating the role of these new immunotherapy agents. These trials include: CheckMate 214 (nivolumab + ipilimumab versus sunitinib in the first-line setting; NCT02231749) and CheckMate 025 (nivolumab versus everolimus after the failure of anti-VEGF treatment; NCT01668784). Interestingly, the primary endpoint (PFS) in the CheckMate 214 trial is primarily powered on patients with intermediate/poor risk and patients are only required to have at least a clear-cell component in their tumour. Therefore, this study will enrol patients that are usually underrepresented in RCC trials. Two large randomised phase III trials investigating the benefit of the anti-tumour vaccine (IMA901; NCT01265901) or autologous dendritic cells (AGS-003; NCT01582672) in addition to sunitinib are expected to be reported in 2015 and 2016, respectively.

References

1 Escudier B, Kataja V: Renal cell carcinoma : ESMO clinical practice guidelines for diagnosis, treatment and follow-up clinical practice guidelines. Ann Oncol 2014;25(suppl 3):49–56.

2 Atzpodien J, Schmitt E, Gertenbach U, Fornara P, Heynemann H, Maskow A, et al: Adjuvant treatment with interleukin-2- and interferon-alpha2a-based chemoimmunotherapy in renal cell carcinoma post tumour nephrectomy: results of a prospectively randomised trial of the German Cooperative Renal Carcinoma Chemoimmunotherapy Group (DGCIN). Br J Cancer 2005;92:843–846.

3 Steineck G, Strander H, Carbin BE, Borgström E, Wallin L, Achtnich U, et al: Recombinant leukocyte interferon alpha-2a and medroxyprogesterone in advanced renal cell carcinoma: a randomized trial. Acta Oncol 1990;29:155–162.

4 Pyrhönen S, Salminen E, Ruutu M, Lehtonen T, Nurmi M, Tammela T, et al: Prospective randomized trial of interferon alfa-2a plus vinblastine versus vinblastine alone in patients with advanced renal cell cancer. J Clin Oncol 1999;17:2859–2867.

5 Coppin C, Porzsolt F, Awa A, Kumpf J, Coldman A, Wilt T: Immunotherapy for advanced renal cell cancer. Cochrane Database Syst Rev 2005:CD001425.

6 Yang JC, Sherry RM, Steinberg SM, Topalian SL, Schwartzentruber DJ, Hwu P, et al: Randomized study of high-dose and low-dose interleukin-2 in patients with metastatic renal cancer. J Clin Oncol 2003;21:3127–3132.

7 McDermott DF, Regan MM, Clark JI, Flaherty LE, Weiss GR, Logan TF, et al: Randomized phase III trial of high-dose interleukin-2 versus subcutaneous interleukin-2 and interferon in patients with metastatic renal cell carcinoma. J Clin Oncol 2005;23:133–141.

8 Fyfe G, Fisher RI, Rosenberg SA, Sznol M, Parkinson DR, Louie AC: Results of treatment of 255 patients with metastatic renal cell carcinoma who received high-dose recombinant interleukin-2 therapy. J Clin Oncol 1995;13:688–696.

9 Fisher RI, Rosenberg SA, Fyfe G: Long-term survival update for high-dose recombinant interleukin-2 in patients with renal cell carcinoma. Cancer J Sci Am 2000;6(suppl 1):55–57.

10 Gore ME, Griffin CL, Hancock B, Patel PM, Pyle L, Aitchison M, et al: Interferon alfa-2a versus combination therapy with interferon alfa-2a, interleukin-2, and fluorouracil in patients with untreated metastatic renal cell carcinoma (MRC RE04/EORTC GU 30012): an open-label randomised trial. Lancet 2010;375:641–648.

11 Escudier B, Pluzanska A, Koralewski P, Ravaud A, Bracarda S, Szczylik C, et al: Bevacizumab plus interferon alfa-2a for treatment of metastatic renal cell carcinoma: a randomised, double-blind phase III trial. Lancet 2007;370:2103–2111.

12 Rini BI, Halabi S, Rosenberg JE, Stadler WM, Vaena DA, Ou SS, et al: Bevacizumab plus interferon alfa compared with interferon alfa monotherapy in patients with metastatic renal cell carcinoma: CALGB 90206. J Clin Oncol 2008;26:5422–5428.

13 Hodi FS, O'Day SJ, McDermott DF, Weber RW, Sosman JA, Haanen JB, et al: Improved survival with ipilimumab in patients with metastatic melanoma. N Engl J Med 2010;363:711–723.

14 Robert C, Ribas A, Wolchok JD, Hodi FS, Hamid O, Kefford R, et al: Anti-programmed-death-receptor-1 treatment with pembrolizumab in ipilimumab-refractory advanced melanoma: a randomised dose-comparison cohort of a phase 1 trial. Lancet 2014;384:1109–1117.

15 Robert C, Thomas L, Bondarenko I, O'Day S, Weber J, Garbe C, et al: Ipilimumab plus dacarbazine for previously untreated metastatic melanoma. N Engl J Med 2011;364:2517–2526.

16 Robert C, Long GV, Brady B, Dutriaux C, Maio M, Mortier L, et al: Nivolumab in previously untreated melanoma without BRAF mutation. N Engl J Med 2015;372:320–330.

17 Ansell SM, Lesokhin AM, Borrello I, Halwani A, Scott EC, Gutierrez M, et al: PD-1 blockade with nivolumab in relapsed or refractory Hodgkin's lymphoma. N Engl J Med 2015;372:311–319.

18 FDA: FDA expands approved use of Opdivo to treat lung cancer. 2015. http://www.fda.gov/newsevents/newsroom/pressannouncements/ucm436534.htm.

19 Yang JC, Hughes M, Kammula U, Royal R, Sherry RM, Topalian SL, et al: Ipilimumab (anti-CTLA4 antibody) causes regression of metastatic renal cell cancer associated with enteritis and hypophysitis. J Immunother 2007;30:825–830.

20 Rini BI, Stein M, Shannon P, Eddy S, Tyler A, Stephenson JJ, et al: Phase 1 dose-escalation trial of tremelimumab plus sunitinib in patients with metastatic renal cell carcinoma. Cancer 2011;117:758–767.

21 Topalian SL, Hodi FS, Brahmer JR, Gettinger SN, Smith DC, McDermott DF, et al: Safety, activity, and immune correlates of anti-PD-1 antibody in cancer. N Engl J Med 2012;366:2443–2454.

22 Drake CG, McDermott DF, Sznol M, Choueiri TK, Kluger HM, Powderly JD, et al: Survival, safety, and response duration results of nivolumab (anti-PD-1; BMS-936558; ONO-4538) in a phase I trial in patients with previously treated metastatic renal cell carcinoma (mRCC): long-term patient follow-up (abstract). J Clin Oncol 2013;31(suppl): 4514.
23 Motzer RJ, Rini BI, McDermott DF, Redman BG, Kuzel T, Harrison MR, et al: Nivolumab for metastatic renal cell carcinoma (mRCC): results of a randomized, dose-ranging phase II trial (abstract). J Clin Oncol 2014;32(suppl 15):5009.
24 Cho DC, Sosman JA, Sznol M, Gordon MS, Hollbecque A, Hamid O, et al: Clinical activity, safety, and biomarkers of MPDL3280A, an engineered PD-L1 antibody in patients with metastatic renal cell carcinoma (mRCC) [abstract]. J Clin Oncol 2013; 31(suppl):4505.
25 Lieu C, Bendell J, Powderly JD, Pishvaian MJ, Hochster H, Eckhardt SG, et al: Safety and efficacy, of MPDL3280A (anti-PDL1) in combination with bevacizumab (BEV) and/or chemotherapy (CHEMO) in patients (PTS) with locally advanced or metastastic solid tumours. Ann Onc 2014;25: iv361–iv372.
26 Brahmer JR, Tykodi SS, Chow LQM, Hwu W-J, Topalian SL, Hwu P, et al: Safety and activity of anti-PD-L1 antibody in patients with advanced cancer. N Engl J Med 2012;366:2455–2465.
27 Thompson RH, Kuntz SM, Leibovich BC, Dong H, Lohse CM, Webster WS, et al: Tumor B7-H1 is associated with poor prognosis in renal cell carcinoma patients with long-term follow-up. Cancer Res 2006; 66:3381–3385.
28 Choueiri TK, Figueroa DJ, Fay AP, Signoretti S, Liu Y, Gagnon R, et al: Correlation of PD-L1 tumor expression and treatment outcomes in patients with renal cell carcinoma receiving sunitinib or pazopanib: results from COMPARZ, a randomized controlled trial. Clin Cancer Res 2015;21:1071–1077.
29 Choueiri TK, Fay AP, Gray KP, Callea M, Ho TH, Albiges L, et al: PD-L1 expression in nonclear-cell renal cell carcinoma. Ann Oncol 2014;25:2178–2184.
30 Hammers H, Plimack ER, Infante JR, Ernstoff B, Rini BI, McDermott DF, et al: Phase 1 study of nivolumab in combination with ipilimumab in metastatic renal cell carcinoma (MRCC). ASCO Meeting Abstracts. Ann Oncol 2014;25(suppl 4):4504.
31 Amin A, Plimack ER, Infante JR, Ernstoff MS, Rini BI, McDermott DF, et al: Nivolumab (anti-PD-1; BMS-936558, ONO-4538) in combination with sunitinib or pazopanib in patients (pts) with metastatic renal cell carcinoma (mRCC) [abstract]. J Clin Oncol 2014;32(suppl):1–15.
32 Oudard S, Rixe O, Beuselinck B, Linassier C, Banu E, Machiels J-P, et al: A phase II study of the cancer vaccine TG4010 alone and in combination with cytokines in patients with metastatic renal clear-cell carcinoma: clinical and immunological findings. Cancer Immunol Immunother 2011;60:261–271.
33 Amato RJ, Hawkins RE, Kaufman HL, Thompson JA, Tomczak P, Szczylik C, et al: Vaccination of metastatic renal cancer patients with MVA-5T4: a randomized, double-blind, placebo-controlled phase III study. Clin Cancer Res 2010;16:5539–5547.
34 Adler A, Gillon G, Lurie H, Shaham J, Loven D, Shachter Y, et al: Active specific immunotherapy of renal cell carcinoma patients: a prospective randomized study of hormono-immuno-versus hormonotherapy: preliminary report of immunological and clinical aspects. J Biol Response Mod 1987;6:610–624.
35 Galligioni E, Quaia M, Merlo A, Carbone A, Spada A, Favaro D, et al: Adjuvant immunotherapy treatment of renal carcinoma patients with autologous tumor cells and Bacillus Calmette-Guerin: five-year results of a prospective randomized study. Cancer 1996;77:2560–2566.
36 Jocham D, Richter A, Hoffmann L, Iwig K, Fahlenkamp D, Zakrzewski G, et al: Adjuvant autologous renal tumour cell vaccine and risk of tumour progression in patients with renal-cell carcinoma after radical nephrectomy: phase III, randomised controlled trial. Lancet 2004;363:594–599.
37 Wood C, Srivastava P, Bukowski R, Lacombe L, Gorelov AI, Gorelov S, et al: An adjuvant autologous therapeutic vaccine (HSPPC-96; vitespen) versus observation alone for patients at high risk of recurrence after nephrectomy for renal cell carcinoma: a multicentre, open-label, randomised phase III trial. Lancet 2008;372:145–154.
38 Pizzocaro G, Piva L, Colavita M, Ferri S, Artusi R, Boracchi P, et al: Interferon adjuvant to radical nephrectomy in Robson stages II and III renal cell carcinoma: a multicentric randomized study. J Clin Oncol 2001;19:425–431.
39 Messing EM, Manola J, Wilding G, Propert K, Fleischmann J, Crawford ED, et al: Phase III study of interferon alfa-NL as adjuvant treatment for resectable renal cell carcinoma: an Eastern Cooperative Oncology Group/Intergroup Trial. J Clin Oncol 2003;21: 1214–1222.

40 Clark JI, Atkins MB, Urba WJ, Creech S, Figlin RA, Dutcher JP, et al: Adjuvant high-dose bolus interleukin-2 for patients with high-risk renal cell carcinoma: a cytokine working group randomized trial. J Clin Oncol 2003;21:3133–3140.

41 Passalacqua R, Buzio C, Buti S, Labianca R, Porta C, Boni C, et al: Adjuvant low-dose interleukin-2 (IL-2) plus interferon-alpha (IFN-α) in operable renal cell cancer (RCC): a phase III, randomized, multicenter, independent trial of the Italian Oncology Group for Clinical Research (GOIRC). J Immunother 2014;37: 440–447.

42 Aitchison M, Bray CA, van Poppel H, Sylvester R, Graham J, Innes C, et al: Adjuvant 5-flurouracil, alpha-interferon and interleukin-2 versus observation in patients at high risk of recurrence after nephrectomy for renal cell carcinoma: results of a phase III randomised European Organisation for Research and Treatment of Cancer. Eur J Cancer 2014;50:70–77.

Prof. Martin Gore, PhD, FRCP
Medical Oncology
Royal Marsden Hospital
Fulham Road, SW3 6JJ London (UK)
E-Mail martin.gore@rmh.nhs.uk

Michielin O, Coukos G (eds): Immuno-Oncology.
Prog Tumor Res. Basel, Karger, 2015, vol 42, pp 11–21 (DOI: 10.1159/000436986)

Immunotherapy of Brain Tumors

Valérie Dutoit[a] · Denis Migliorini[b] · Paul R. Walker[a] · Pierre-Yves Dietrich[b]

[a]Laboratory of Tumor Immunology, [b]Center of Oncology, Geneva University Hospitals, Geneva, Switzerland

Abstract

Glioma is one of the most devastating cancers, affecting children and young adults, and associated with a very high morbidity and poor prognosis. The propensity of glioma cells to invade normal brain structures makes current treatments poorly efficient and new therapeutic strategies an urgent need. We now know that many of the rules governing immune responses in other tissues are also valid for the brain, providing solid scientific background for developing new strategies exploiting the immune system to fight brain tumors. Some vaccines use tumor-specific mutated peptides (EGFRvIII, IDH1 or personalized peptides), but most are tumor-associated or undefined tumor-derived peptides (tumor-eluted peptides, peptides predicted from tumor-associated proteins or bound to HSPPC-96 complexes), in some cases pulsed on dendritic cells. Cell therapy is less advanced but the first clinical trials exploring the safety of T cells with chimeric antigen receptors incorporating antibodies to EGFRvIII, IL-13Rα2 or Her2 are ongoing. Finally, various strategies designed at reshaping the glioma microenvironment are being tested, including TGFβ inhibition, Treg depletion and immune checkpoint blockade. Altogether, combining vaccines, cell therapy and reshaping of the tumor microenvironment will be the foundation for a new era of therapeutics for brain tumors.

Primary brain tumors are a heterogeneous group of tumors occurring in the central nervous system. They include less aggressive tumors such as meningiomas and schwannomas, but also tumors with a dismal prognosis, high-grade glioma, of which WHO grade IV astrocytoma or glioblastoma multiforme (GBM) is the most lethal [1]. GBM can present de novo in patients, without evidence of evolution from a lower-grade tumor (primary GBM, ~90% of cases) or as progression from a lower-grade glioma (secondary GBM), with both entities harboring a specific genetic profile [2]. Current treatments for GBM include surgery followed by a concomitant course of radiotherapy and chemotherapy with the drug temozolomide, which is then given alone on a monthly basis as an adjuvant. This treatment, although having shown to prolong patient survival [3], is still poorly efficient, mostly due to treatment-resistant GBM cells, and their high capacity to spread and invade normal brain structures very early in the disease. As a consequence, substantial efforts have been made in the last 20 years to develop new therapeutic options for GBM. Among the various approaches investigated, immunotherapy has emerged as a very promising line of attack [4].

Immunotherapy for Glioma

Tumor immunotherapy, defined as use of the immune system to fight cancer, is considered to be a powerful tool for the treatment of many malignancies. It started to be developed many decades ago for melanoma and has since extended to almost all tumor types [5]. This research has highlighted the rules governing antitumor immune responses and how to exploit them to build immunotherapy. It made clear that tumor growth was recognized by immune sentinel cells as an inflammatory process, initiating innate and adaptive immune responses that could be detected in patients. Adaptive immune responses were mostly identified as tumor-specific T cells in the tumor microenvironment and their presence correlated with increased survival [6]. Regarding T cell immune responses, we learned that tumor cell antigens are captured by local dendritic cells (DC) that migrate to regional lymph nodes to present tumor-derived peptides to T lymphocytes [7]. The latter become activated and migrate back to the tumor site to achieve tumor cell destruction. Although this was clearly demonstrated, results also showed that the tumor was able to implement many mechanisms to counteract the immune response. This resulted, in the majority of cases, in inefficient tumor eradication [8].

Regarding tumors located in the brain, it was long believed that these rules could not apply, the brain being an immunoprivileged organ, not allowing free diffusion of cells and molecules, having no resident DC and lacking conventional lymphatic vessels [9]. However, in the last 20 years it was shown that brain tumors are able to elicit potent antitumor immune responses. Using animal models of intracranial tumors, studies showed that tumor-derived antigens are transported to the cervical lymph nodes to stimulate specific T cells [10]. After amplification, these T cells are able to efficiently migrate back to the brain and destroy tumor cells [11]. Therefore, although we are still lacking precise information about the identity of the antigen-presenting cells that capture tumor-derived antigens in the brain, the general rules of antitumor response can be applied to the brain, providing a solid scientific background for the further development of brain tumor immunotherapy.

Immunotherapy: The Means

There are two major approaches being currently explored for cancer immunotherapy: tumor vaccines and T cell therapy. Both rely on the elicitation of significant numbers of tumor-specific T cells that will patrol the patient's body to seek and destroy tumor cells. Tumor vaccines are therapeutic vaccines that aim at eliciting an immune response in the patient by injection of tumor-derived antigens. It is called active immunotherapy as the patient's immune system plays an active role in the process. T cell therapy, on the contrary, requires the generation of a high number of tumor-specific T cells outside the patient's body (passive immunotherapy), a process that has the unique advantage of allowing optimization of the cell product not only in terms of tumor specificity but also of effector cell migration to the tumor site, persistence in vivo and resistance to the immunosuppressive tumor environment [12].

Cancer Vaccines

Therapeutic cancer vaccines are based on defined tumor-derived antigens or on the tumor cell itself. The main requirement for the formulation of a safe tumor vaccine is the choice of antigen, which has to be present on tumor cells but not on normal cells of the body, or in very small amounts. In this regard, the best antigens are tumor-specific antigens (TSA) arising from mutations in the tumor, as they are absent from normal tissues. Unfortunately, such antigens are usually not shared by the majority of patients. Tumor-associated antigens (TAA), which are not tumor specific but are overexpressed in the tumor as compared to normal tissues, are a second type of antigens that can be used in tumor vaccines. Although their choice has to be very carefully made in order to avoid collateral tissue damage, these antigens are shared by a larger proportion of patients. Their relatively high frequency also allows designing multiantigen vaccines, therefore reducing the risk for tumor escape. Apart from these well-defined antigens, whole tumor cells or tumor-derived undefined antigens (tumor mRNA, eluted peptides, peptides bound to heat shock proteins) are used. Although these antigens are available for any given patient, their undefined nature renders them potentially harmful as they contain a nontumor-specific component. Additionally, they do not allow precise monitoring of the induced immune response and results from trials using these antigens are therefore difficult to interpret.

Past and Ongoing Vaccine Trials

TSA-Based Vaccines. There have been encouraging clinical trials with the prototype of TSA in GBM, the EGFRvIII protein, which is generated by exon splicing of the EGFR protein and is solely expressed by tumor cells [13]. These studies revealed that cellular and humoral immune responses are elicited after vaccination with a 14-amino acid peptide spanning the mutation coupled to keyhole limpet hemocyanin that correlated with improved outcome [14, 15]. Interestingly, the majority of patients that relapsed after vaccination had lost the EGFRvIII antigen [15], demonstrating at the same time proof-of-principle of efficacy of vaccine-induced immune responses in eradicating tumor cells but also tumor escape from immune pressure when a single antigen is used. The EGFRvIII vaccine (Rindopepimut) is now in phase III clinical trial in newly diagnosed (NCT01480479; table 1) and relapsed (NCT01498328) GBM. Unfortunately, as this mutation is present in only about 20% of GBM [16], its use cannot be extended to every glioma patient. Another TSA being tested in a clinical trial for recurrent grade II astrocytoma (NCT02193347) is based on a mutation in the IDH1 protein [17], which is present on low-grade glioma and secondary GBM [2]. Again, this mutated antigen represents a novel target for glioma, with the limitation that it is present in a subset of patients only. Altogether, the recognition that individual TSA are not widely distributed but are more patient specific has led to the development of personalized TSA-based vaccines in glioma (NCT02149225) [18] and other malignancies [19]. The principle of this approach is to sequence the patient's tumor to detect individual mutations. Mutations are then compared to eluted peptides from the tumor cell surface to ensure the selection of mutated peptides that are presented at the tumor cell surface. This second step, which is undertaken in the glioma trial mentioned above but usually overlooked in other trials, is a prerequisite for tumor eradication, as very few mutations are indeed presented at the cell surface as peptides and, therefore, eliciting immune responses to any peptide spanning a mutation will in the majority of cases offer no clinical benefit as elicited T cells will not find their target.

TAA-Based Vaccines. Therapeutic vaccine trials exploring defined TAA in GBM have used multiple antigens in order to have broad tumor targeting and avoid tumor escape. These trials, using peptide-pulsed DC or peptides alone in adjuvant, demonstrated that T cell responses are elicited in patients with no sign of collateral autoimmunity and showed benefit in some patients [20–22], including pediatric patients [23]. These results have prompted initiation of more clinical trials, extending the patient population to lower-grade glioma, oligoastrocytoma and oligodendroglioma (results awaited) and ependymoma (NCT01795313). Recently, a novel set of glioma-associated TAA became available to be used as a peptide vaccine. Peptide elution from GBM cells was used to identify 10 novel GBM-associated antigens, a process that ensured that identified antigens are presented at the tumor cell surface and thus visible to the immune system [24]. These antigens are overexpressed in 80–100% of patients and one of them induced a spontaneous antiglioma T cell response, as evidenced by the detection of tumor-infiltrating lymphocytes at the tumor site [24]. Three trials (NCT01403285, NCT01222221 and NCT01920191) incorporating these well-characterized TAA (called the IMA950 antigens) are ongoing using different adjuvants (GM-CSF, GM-CSF/imiquimod and poly-ICLC). Interestingly, individually defined well-characterized tumor-eluted peptides are being used in addition to mutated peptides in the personalized glioma vaccine trial mentioned above (NCT02149225). Finally, whereas the majority of the above-mentioned peptide vaccines contain only

Table 1. Ongoing immunotherapy trials in brain tumors

Tumor vaccines	Antigen format	NCT number
TSA		
1. EGFRvIII	14-mer peptide spanning the mutation	NCT01480479 NCT01498328 NCT01967758
2. IDH1	25-mer peptide spanning the mutation	NCT02193347
3. Personalized mutated peptides	9- to 19-mer peptides	NCT02149225
TAA		
4. EphA2, IL-13Rα2, YKL-40, gp100	A2-restricted peptides	NCT01795313
5. BCAN, CSPG4, FABP7, IGF2BP3, NLGN4X, NRCAM, PTPRZ1, TNC, survivin, c-met	A2-restricted peptides + 2 CD4 epitopes	NCT01403285 NCT01222221 NCT01920191
6. CD133	A2-restricted peptide	NCT02049489
7. Survivin	A2/A3-restricted peptides	NCT01250470
8. Individually tumor-eluted peptides	A2/A24-restricted peptides	NCT02149225
9. NY-ESO-1	protein fused to a DC targeting antibody	NCT01522820
10. CMV pp65	whole protein mRNA	NCT00626483
11. CEA	protein-expressing virus	NCT00390299
Undefined antigens (tumor-based vaccines)		
12. Tumor lysate		NCT01808820 NCT01902771 NCT01204684 NCT01635283 NCT01792505 NCT00045968
13. HSPPC-96-complexed peptides		NCT01814813 NCT02122822
14. Stem cell line		NCT01400672 NCT01678352 NCT02010606
CARs		
15. EGFRvIII		NCT01454596 NCT02209376
16. Her2		NCT01109095
17. IL-13Rα chain		NCT02208362

Table 1. Continued

Immunomodulation	Other intervention	
Treg cells		
18. Anti-CD25 antibody	CMV-mRNA vaccine	NCT00626483
19. Cyclophosphamide	EGFRvIII CAR, TAA vaccine, tumor lysate vaccines	NCT01454596 NCT01403285 NCT01903330
VEGF		
20. Anti-VEGF antibody	EGFRvIII vaccine, stem cell line vaccine, checkpoint blockade antibodies	NCT01498328 NCT02010606 NCT02017717
STAT3		
21. Stat3 inhibitor	none	NCT01904123
IDO		
22. IDO inhibitor (1-Me-tryptophan)	none	NCT02052648
Immune checkpoint inhibitors		
23. Anti-CTLA4 antibody	anti-PD1 antibody	NCT02017717
24. Anti-PD1 antibody	none or anti-CTLA4 antibody	NCT01952769 NCT02017717

BCAN = Brevican; CSPG4 = chondroitin sulfate proteoglycan 4; EGFRvIII = epidermal growth factor receptor mutant form lacking exons 2–8; EphA2 = ephrin type-A receptor 2; FABP7 = fatty acid binding protein 7; brain; gp100 = glycoprotein 100 or melanocyte protein PMEL; Her2 = human epidermal growth factor receptor-2; HSPPC-96 = heat shock protein peptide complex-96; IGF2BP3 = insulin-like growth factor 2 mRNA-binding protein 3; IL-13Rα = interleukin-13 receptor subunit alpha-2; NLGN4X = neuroligin-4, X-linked; NRCAM = neuronal cell adhesion molecule; PTPRZ1 = protein tyrosine phosphatase receptor-type zeta polypeptide 1; TNC = tenascin C; YKL-40 = chitinase-3-like protein 1.

CD8 T cell epitopes, some TAA vaccines use whole proteins as immunogen, with the goal of eliciting both CD4 and CD8 T cell responses (NCT01522820, NCT00626483, NCT00390299).

Tumor-Based Vaccines. DC pulsed with tumor cell lysates, eluted peptides or HSPPC-96-complexed peptides have also been investigated in clinical trials and have shown the feasibility and safety of the approach, without any autoimmune reaction being observed [4]. Although immune monitoring of these trials is difficult and limited due to the undefined nature of the antigens, some studies have shown humoral and/or cellular immune responses to the immunogen. The survival benefit of these vaccines, however, is still to be determined. Nevertheless, more studies are ongoing, some using glioma stem cell lines (table 1). Glioma stem cells are a subpopulation of tumor cells that participate in tumor initiation and propagation and are relatively resistant to radio- and chemotherapy [25], making their targeting attractive for proper tumor elimination.

Vaccine Adjuvants. The vast majority of glioma and other tumor vaccines are injected together with an immune stimulator or immune adjuvant. These adjuvants play a major role in augmenting vaccine efficacy through stimulation of the innate immune response, DC maturation or modulation of T regulatory cell activity. In glioma trials, Montanide has been used to create an antigen depot [20, 23], GM-CSF has been used with the EGFRvIII vaccine to recruit and enhance the function of DCs [26] and the TLR (Toll-like receptor) 7 agonist imiquimod has been occasionally used [27]. The TLR3 agonist poly-ICLC (polyinosinic-polycytidylic acid stabilized with poly L-lysine and carboxymethyl cellulose) has been frequently and safely explored in glioma and showed particularly interesting properties in terms of inducing a proinflammatory response and helping vaccine-elicited T cells migrating to the brain [20, 23, 28]. GM-CSF, poly-ICLC and imiquimod are being used in current trials; however, how best to formulate glioma vaccines will require a systematic comparison of these different vaccine formulations.

T Cell Therapy

Adoptive T cell transfer has evolved remarkably in the last 20 years with the development of genetic engineering allowing the generation of high avidity tumor-specific T cell products. Tumor specificity can be achieved through expression of a well-characterized tumor-reactive T cell receptor (TCR engineering) or of a tumor-binding antibody, generating chimeric antigen receptor (CAR) T cells [12]. Although TCR engineering is not yet being developed for glioma, CARs targeting several proteins (IL-13 receptor [29, 30], EphA2 [31], Her2 [32, 33] and EGFRvIII [34–36]) have shown promising preclinical results. Clinical trials are currently ongoing for IL-13Rα2 (NCT02208362), Her2 (NCT01109095) and EGFRvIII (NCT01454596 and NCT02209376), and will reveal whether therapeutic benefit can be achieved without unacceptable toxicity.

Further Strategies to Improve Immunotherapy

Immunotherapy alone will most probably not be sufficient for the successful eradication of brain tumors. The current view in the field, although not yet implemented in practice, is that efficacy will be improved by determining synergistic combinations of immunotherapy with chemotherapy, radiotherapy, small molecules inhibitors, or strategies to target the immunosuppressive tumor microenvironment.

Modifying the Local Tumor Environment

Glioma, similar to other tumors, is a master in developing strategies to subvert immune responses, both at the cellular and molecular levels [37]. T regulatory (Treg) cells are known to be recruited at the tumor site very early during antiglioma immune responses [38] and inhibit cytotoxic T cells. Many clinical trials are therefore ongoing with the aim of depleting Tregs during immunotherapy. Studies using an anti-CD25 (the IL-2Rα chain) have shown that Treg depletion was well tolerated and did not prevent vaccine-induced immune responses in the lymphodepleted environment found in GBM patients [39]. A clinical trial using the anti-CD25 antibody during vaccination with RNA-loaded DC is ongoing (NCT00626483). Cyclophosphamide, an alkylating agent, has been shown to decrease Treg numbers and function [40], and low-dose cyclophosphamide is being used in vaccine trials for many malignancies to suppress Tregs. In glioma, a single dose of cyclophosphamide is being given prior to the IMA950 peptide vaccine (NCT01403285), tumor-lysate pulsed DC vaccines (NCT01903330) or to EGFRvIII-directed CAR T cell transfer (NCT01454596). In addition to promoting Treg infiltration and function, the glioma tumor environment is able to convert macrophages from a pro- to an anti-inflammatory phenotype and monocytes to MDSCs (myeloid-derived suppressor cells) [41], two mechanisms that can suppress T cell function. These are not yet targeted in glioma immunotherapy trials but should receive attention in the future.

Among molecular mechanisms of T cell suppression in glioma, TGF-β2 is probably the most relevant and, although initial trials have been disappointing, it is actively targeted in glioma with TGF-β2 antisense nucleotide or TGF-β RI kinase inhibitors (results awaited). VEGF (vascular endothelial growth factor) also contributes to immunosuppression in the tumor microenvironment through inhibition of DC maturation, promotion of MDSCs and inhibition of T cell migration in the brain [42, 43]. The anti-VEGF antibody bevacizumab is routinely being used for recurrent GBM [44] and is being tested in combination with the EGFRvIII peptide vaccine in patients with recurrent GBM (NCT01498328) or with stem cell line vaccination in newly diagnosed or recurrent GBM (NCT02010606). STAT3 (signal transducer and activator of transcription 3) is a transcription factor downstream of many signaling pathways (EGFR, c-met, etc.) that is constitutively active in glioma and is therefore persistently phosphorylated in these tumors [45]. A STAT3 inhibitor is currently being tested as a single drug after standard-of-care treatment in patients with glioma (NCT01904123). Finally, the enzyme indoleamine 2,3 dioxygenase (IDO), which promotes immunosuppression by depleting tryptophan in the tumor microenvironment, is being targeted using the IDO inhibitor 1-methyl-tryptophan (NCT02052648) in recurrent GBM.

Immune Checkpoint Blockade

Immune checkpoint molecules are expressed during normal T cell responses to prevent overactivation of the immune system. Although the list is growing fast, the two most studied molecules are cytotoxic T lymphocyte-associated protein 4 (CTLA-4) and programmed cell death-1 (PD-1) [46]. Whereas CTLA-4 acts early in T cell activation, PD-1 blocks T cells at later stages of the immune response in the local sites of inflammation. Consistent with this, CTLA-4 and PD-1 are highly expressed on T cells subjected to chronic antigen exposure and on tumor-infiltrating lymphocytes, a phenotype associated with reduced effector functions [46]. In addition, PD-1 ligands PD-L1 and

PD-L2 are frequently upregulated in the tumor environment, and this is also the case for PD-L1 in glioma [47]. Recent clinical trials in metastatic melanoma have shown the dramatic effect of anti-CTLA-4 and anti-PD-1/PD-L1 antibodies on improving patient survival [48] and the CTLA-4 antibody ipilimumab was approved in 2011 by the FDA for this indication. Anti-PD-1/PD-L1 antibodies seem to be even more efficient with less toxicity and are currently being tested for many malignancies. This led to approval of the anti-PD-L1 antibody pembrolizumab for the treatment of metastatic melanoma by the FDA in 2014. In glioma, the anti-PD-1 antibody nivolumab is being tested in recurrent GBM alone or in combination with ipilimumab, and is also being compared with bevacizumab (NCT02017717). Another PD-1 antibody, pidilizumab, is being tested for diffuse intrinsic pontine glioma and recurrent GBM (NCT01952769). Although these antibodies have thus far been tested as individual drugs in glioma, their mechanism of action is mostly through enhancing antitumor immune responses, and adding them to immunotherapeutic strategies is likely to be synergistic.

Conclusions

Manipulating the immune system to achieve an antitumor effect is a fascinating field still in the process of being fully developed. Many approaches are being explored and have already shown highly encouraging results. The relative ease of application of T cell vaccines and the potential for optimization offered by T cell therapy open many opportunities for future glioma patient treatment. Adding strategies to resist the detrimental tumor microenvironment should be incorporated for generating a powerful therapy for brain tumor patients.

References

1 Woodworth GF, Dunn GP, Nance EA, Hanes J, Brem H: Emerging insights into barriers to effective brain tumor therapeutics. Front Oncol 2014;4:126.

2 Ohgaki H, Kleihues P: The definition of primary and secondary glioblastoma. Clin Cancer Res 2013;19: 764–772.

3 Stupp R, Mason WP, van den Bent MJ, Weller M, Fisher B, Taphoorn MJ, et al: Radiotherapy plus concomitant and adjuvant temozolomide for glioblastoma. N Engl J Med 2005;352:987–996.

4 Reardon DA, Wucherpfennig KW, Freeman G, Wu CJ, Chiocca EA, Wen PY, et al: An update on vaccine therapy and other immunotherapeutic approaches for glioblastoma. Expert Rev Vaccines 2013;12:597–615.

5 Lizee G, Overwijk WW, Radvanyi L, Gao J, Sharma P, Hwu P: Harnessing the power of the immune system to target cancer. Annu Rev Med 2013;64:71–90.

6 Fridman WH, Pages F, Sautes-Fridman C, Galon J: The immune contexture in human tumours: impact on clinical outcome. Nat Rev Cancer 2012;12:298–306.

7 Mellman I, Coukos G, Dranoff G: Cancer immunotherapy comes of age. Nature 2011;480:480–489.

8 Schreiber RD, Old LJ, Smyth MJ: Cancer immunoediting: integrating immunity's roles in cancer suppression and promotion. Science 2011;331: 1565–1570.

9 Carson MJ, Doose JM, Melchior B, Schmid CD, Ploix CC: CNS immune privilege: hiding in plain sight. Immunol Rev 2006;213:48–65.

10 Calzascia T, Masson F, Berardino-Besson W, Contassot E, Wilmotte R, Aurrand-Lions M, et al: Homing phenotypes of tumor-specific CD8 T cells are predetermined at the tumor site by crosspresenting APCs. Immunity 2005;22:175–184.

11 Masson F, Calzascia T, Berardino-Besson W, de Tribolet N, Dietrich PY, Walker PR: Brain microenvironment promotes the final functional maturation of tumor-specific effector CD8+ T cells. J Immunol 2007;179:845–853.

12 Restifo NP, Dudley ME, Rosenberg SA: Adoptive immunotherapy for cancer: harnessing the T cell response. Nat Rev Immunol 2012;12:269–281.

13 Sampson JH, Archer GE, Mitchell DA, Heimberger AB, Bigner DD: Tumor-specific immunotherapy targeting the EGFRvIII mutation in patients with malignant glioma. Semin Immunol 2008;20:267–275.

14 Sampson JH, Archer GE, Mitchell DA, Heimberger AB, Herndon JE, Lally-Goss D, et al: An epidermal growth factor receptor variant III-targeted vaccine is safe and immunogenic in patients with glioblastoma multiforme. Mol Cancer Ther 2009;8:2773–2779.

15 Sampson JH, Heimberger AB, Archer GE, Aldape KD, Friedman AH, Friedman HS, et al: Immunologic escape after prolonged progression-free survival with epidermal growth factor receptor variant III peptide vaccination in patients with newly diagnosed glioblastoma. J Clin Oncol 2010;28:4722–4729.

16 Choi BD, Archer GE, Mitchell DA, Heimberger AB, McLendon RE, Bigner DD, et al: EGFRvIII-targeted vaccination therapy of malignant glioma. Brain Pathol 2009;19:713–723.

17 Schumacher T, Bunse L, Pusch S, Sahm F, Wiestler B, Quandt J, et al: A vaccine targeting mutant IDH1 induces antitumour immunity. Nature 2014;512:324–327.

18 Rammensee HG, Singh-Jasuja H: HLA ligandome tumor antigen discovery for personalized vaccine approach. Expert Rev Vaccines 2013;12:1211–1217.

19 Vonderheide RH, Nathanson KL: Immunotherapy at large: the road to personalized cancer vaccines. Nat Med 2013;19:1098–1100.

20 Okada H, Kalinski P, Ueda R, Hoji A, Kohanbash G, Donegan TE, et al: Induction of $CD8^+$ T-cell responses against novel glioma-associated antigen peptides and clinical activity by vaccinations with α-type 1 polarized dendritic cells and polyinosinic-polycytidylic acid stabilized by lysine and carboxymethylcellulose in patients with recurrent malignant glioma. J Clin Oncol 2011;29:330–336.

21 Phuphanich S, Wheeler CJ, Rudnick JD, Mazer M, Wang H, Nuno MA, et al: Phase I trial of a multi-epitope-pulsed dendritic cell vaccine for patients with newly diagnosed glioblastoma. Cancer Immunol Immunother 2013;62:125–135.

22 Terasaki M, Shibui S, Narita Y, Fujimaki T, Aoki T, Kajiwara K, et al: Phase I trial of a personalized peptide vaccine for patients positive for human leukocyte antigen-A24 with recurrent or progressive glioblastoma multiforme. J Clin Oncol 2011;29:337–344.

23 Pollack IF, Jakacki RI, Butterfield LH, Hamilton RL, Panigrahy A, Potter DM, et al: Antigen-specific immune responses and clinical outcome after vaccination with glioma-associated antigen peptides and polyinosinic-polycytidylic acid stabilized by lysine and carboxymethylcellulose in children with newly diagnosed malignant brainstem and nonbrainstem gliomas. J Clin Oncol 2014;32:2050–2058.

24 Dutoit V, Herold-Mende C, Hilf N, Schoor O, Beckhove P, Bucher J, et al: Exploiting the glioblastoma peptidome to discover novel tumour-associated antigens for immunotherapy. Brain 2012;135:1042–1054.

25 Ahmed AU, Auffinger B, Lesniak MS: Understanding glioma stem cells: rationale, clinical relevance and therapeutic strategies. Expert Rev Neurother 2013;13:545–555.

26 Swartz AM, Li QJ, Sampson JH: Rindopepimut: a promising immunotherapeutic for the treatment of glioblastoma multiforme. Immunotherapy 2014;6:679–690.

27 Prins RM, Soto H, Konkankit V, Odesa SK, Eskin A, Yong WH, et al: Gene expression profile correlates with T cell infiltration and survival in glioblastoma patients vaccinated with dendritic cell immunotherapy. Clin Cancer Res 2011;17:1603–1615.

28 Prins RM, Soto H, Konkankit V, Odesa SK, Eskin A, Yong WH, et al: Gene expression profile correlates with T-cell infiltration and relative survival in glioblastoma patients vaccinated with dendritic cell immunotherapy. Clin Cancer Res 2011;17:1603–1615.

29 Kong S, Sengupta S, Tyler B, Bais AJ, Ma Q, Doucette S, et al: Suppression of human glioma xenografts with second-generation IL13R-specific chimeric antigen receptor-modified T cells. Clin Cancer Res 2012;18:5949–5960.

30 Brown CE, Starr R, Aguilar B, Shami AF, Martinez C, D'Apuzzo M, et al: Stem-like tumor-initiating cells isolated from IL13Rα2 expressing gliomas are targeted and killed by IL13-zetakine-redirected T Cells. Clin Cancer Res 2012;18:2199–2209.

31 Chow KK, Naik S, Kakarla S, Brawley VS, Shaffer DR, Yi Z, et al: T cells redirected to EphA2 for the immunotherapy of glioblastoma. Mol Ther 2013;21:629–637.

32 Ahmed N, Salsman VS, Kew Y, Shaffer D, Powell S, Zhang YJ, et al: HER2-specific T cells target primary glioblastoma stem cells and induce regression of autologous experimental tumors. Clin Cancer Res 2010;16:474–485.

33 Hegde M, Corder A, Chow KK, Mukherjee M, Ashoori A, Kew Y, et al: Combinational targeting offsets antigen escape and enhances effector functions of adoptively transferred T cells in glioblastoma. Mol Ther 2013;21:2087–2101.

34 Ohno M, Ohkuri T, Kosaka A, Tanahashi K, June CH, Natsume A, et al: Expression of miR-17-92 enhances anti-tumor activity of T-cells transduced with the anti-EGFRvIII chimeric antigen receptor in mice bearing human GBM xenografts. J Immunother Cancer 2013;1:21.
35 Shen CJ, Yang YX, Han EQ, Cao N, Wang YF, Wang Y, et al: Chimeric antigen receptor containing ICOS signaling domain mediates specific and efficient antitumor effect of T cells against EGFRvIII expressing glioma. J Hematol Oncol 2013;6:33.
36 Sampson JH, Choi BD, Sanchez-Perez L, Suryadevara CM, Snyder DJ, Flores CT, et al: EGFRvIII mCAR-modified T-cell therapy cures mice with established intracerebral glioma and generates host immunity against tumor-antigen loss. Clin Cancer Res 2014;20:972–984.
37 Platten M, Ochs K, Lemke D, Opitz C, Wick W: Microenvironmental clues for glioma immunotherapy. Curr Neurol Neurosci Rep 2014;14:440.
38 Tran Thang NN, Derouazi M, Philippin G, Arcidiaco S, Berardino-Besson W, Masson F, et al: Immune Infiltration of Spontaneous Mouse Astrocytomas Is Dominated by Immunosuppressive Cells from Early Stages of Tumor Development. Cancer Res 2010;70:4829–4839.
39 Sampson JH, Schmittling RJ, Archer GE, Congdon KL, Nair SK, Reap EA, et al: A pilot study of IL-2Rα blockade during lymphopenia depletes regulatory T-cells and correlates with enhanced immunity in patients with glioblastoma. PLoS One 2012;7:e31046.
40 Le DT, Jaffee EM: Regulatory T-cell modulation using cyclophosphamide in vaccine approaches: a current perspective. Cancer Res 2012;72:3439–3444.
41 Kohanbash G, Okada H: Myeloid-derived suppressor cells (MDSCs) in gliomas and glioma-development. Immunol Invest 2012;41:658–679.
42 Huang Y, Goel S, Duda DG, Fukumura D, Jain RK: Vascular normalization as an emerging strategy to enhance cancer immunotherapy. Cancer Res 2013;73:2943–2948.
43 Fricke I, Mirza N, Dupont J, Lockhart C, Jackson A, Lee JH, et al: Vascular endothelial growth factor-trap overcomes defects in dendritic cell differentiation but does not improve antigen-specific immune responses. Clin Cancer Res 2007;13:4840–4848.
44 Chowdhary S, Chamberlain M: Bevacizumab for the treatment of glioblastoma. Expert Rev Neurother 2013;13:937–949.
45 Luwor RB, Stylli SS, Kaye AH: The role of Stat3 in glioblastoma multiforme. J Clin Neurosci 2013;20:907–911.
46 Pardoll DM: The blockade of immune checkpoints in cancer immunotherapy. Nat Rev Cancer 2012;12:252–264.
47 Wilmotte R, Burkhardt K, Kindler V, Belkouch MC, Dussex G, Tribolet N, et al: B7-homolog 1 expression by human glioma: a new mechanism of immune evasion. Neuroreport 2005;16:1081–1085.
48 Ott PA, Hodi FS, Robert C: CTLA-4 and PD-1/PD-L1 blockade: new immunotherapeutic modalities with durable clinical benefit in melanoma patients. Clin Cancer Res 2013;19:5300–5309.

Pierre-Yves Dietrich, MD
Center of Oncology, Geneva University Hospitals
Rue Gabrielle-Perret-Gentil 4
1211 Geneva 14 (Switzerland)
E-Mail pierre-yves.dietrich@hcuge.ch

Michielin O, Coukos G (eds): Immuno-Oncology.
Prog Tumor Res. Basel, Karger, 2015, vol 42, pp 22–29 (DOI: 10.1159/000436998)

Immunotherapy of Melanoma

Alexandra Snyder[a, d] · Dmitriy Zamarin[a, b, d] · Jedd D. Wolchok[a–d]

[a]Department of Medicine, [b]Swim Across America-Ludwig Collaborative Research Laboratory and [c]Immunology Program, Ludwig Center for Cancer Immunotherapy, Memorial Sloan-Kettering Cancer Center, and [d]Weill Cornell Medical College, New York, N.Y., USA

Abstract

The history of immunotherapy is rooted in the treatment of melanoma and therapy with immune checkpoint-blocking agents is now a cornerstone for the treatment of metastatic melanoma. The first effective immunotherapies approved by the US Food and Drug Administration in melanoma included interleukin-2 for metastatic disease and interferon alpha in the adjuvant setting. These were followed by a group of new therapies, including checkpoint-blocking antibodies targeting cytotoxic T lymphocyte-associated protein 4 and programmed cell death protein 1. Therapies intended to 'reeducate' T cells, such as tumor-infiltrating lymphocyte therapy, oncolytic viruses and tumor vaccines, have yielded promising results and are under development. Finally, the integration of the above therapies as well as development of new coinhibitory and costimulatory agents, though in early stages, appear very promising and likely represent the next phase in drug development for the treatment of metastatic melanoma.

Melanoma has represented the frontier for exploration of immunotherapy for over a century, and as such patients with metastatic melanoma have been among the first to reap the benefits of checkpoint blockade therapies. From reports of spontaneous tumor regressions and William B. Coley's use of 'Coley's toxins' in the 1890s, investigators focused on cytokine therapies with interleukin-2 (IL-2) [1] and interferon alpha (IFN-α). In the early 2000s, immunotherapies began to diversify, with recent studies now testing combinations of these therapies.

Cytokines and Biochemotherapy

After mouse models showed that IL-2 could cause tumor regression, this drug was developed in patients with metastatic melanoma [2] and was approved for the treatment of patients with metastatic melanoma in 1998. High-dose bolus regimens led to a complete response in 4–6% of patients, with an objective response in an additional 10–15% [1, 3] at the cost of substantial toxicity [3]. IFN-α is approved for the adjuvant treatment of melanoma, as its efficacy in metastatic disease is limited to patients with small volume disease and is of short duration [4]. In a meta-analysis of 14 randomized trials including more than 8,000 patients treated with IFN-α versus placebo or the inactive vaccine GM2-KLH, disease-free and overall survival were increased [5]. However, a separate pooled analysis of cooperative group studies did not show an improvement in overall survival [6]. Therefore, the role of IFN-α in the adjuvant treatment of melanoma remains an issue of debate. In a phase III trial (S0008), 432 patients at high risk for recurrence were randomized to biochemotherapy with the combination of cisplatin, vinblastine and dacarbazine with IL-2 and IFN-α for 3 cycles (over 9 weeks), or IFN-α for 1 year [7]. Progression-free survival was longer in the biochemotherapy arm (4 versus 1.9 years, HR 0.75, 95% CI 0.58–0.97), but overall survival did not improve, and toxicities were substantial.

Targeting T Cell Coinhibitory Receptors

The targets most advanced in development include cytotoxic T lymphocyte-associated protein 4 (CTLA-4), programmed cell death protein-1 (PD-1) and PD-1 ligand 1 (PD-L1; fig. 1). The anti-CTLA-4 agent ipilimumab was FDA-approved in 2011 for the treatment of metastatic melanoma based on two randomized trials. In the first, patients who had received prior systemic therapy were treated with ipilimumab with or without glycoprotein 100 peptide vaccine (gp100, a melanoma antigen) or gp100 peptide vaccine alone [8]. Overall survival was longer in the ipilimumab-containing arms (10 vs. 6.4 months in gp100 alone), with a subset experiencing long-term disease control. This was the first randomized trial to show an improvement in overall survival for any agent in metastatic melanoma. In the second trial, treatment-naive patients were randomized to dacarbazine with or without ipilimumab [9]. Overall survival was prolonged in the ipilimumab-treated group (11.2 vs. 9.1 months, HR for death 0.72, $p < 0.001$) at the expense of greater toxicity. Phase I and II trials of another anti-CTLA-4 agent, tremelimumab, yielded promising results [10, 11], but no overall survival advantage was seen in a randomized phase III study of tremelimumab versus chemotherapy [12], likely due, in part, to the crossover of patients on the chemotherapy arm to treatment with ipilimumab. In the adjuvant setting, ipilimumab is undergoing study in EORTC 18071, in which 951 patients with stage III melanoma were randomized to ipilimumab versus placebo [13]. Those in the ipilimumab arm experienced an increase in median relapse-free survival of 26 versus 17 months (HR 0.75, 95% CI 0.64–0.9); as of early 2015, survival data have not matured. Notably, 36.5% of patients experienced grade 3 adverse effects, and there were 5 treatment-related deaths.

The anti-PD-1 antibodies for melanoma most advanced in development include nivolumab and pembrolizumab. In a phase I/II study of nivolumab including 107

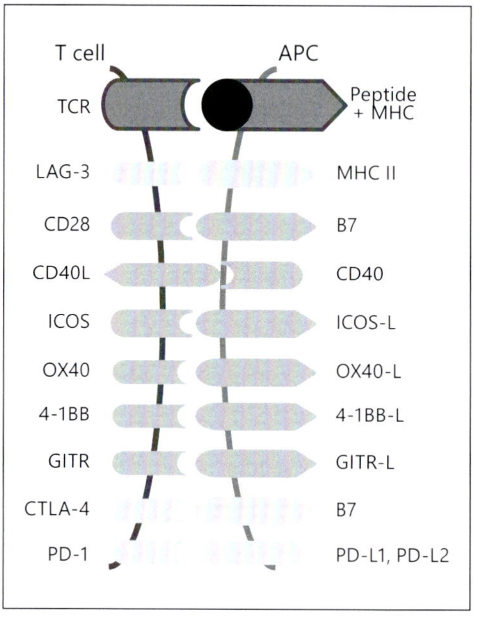

Fig. 1. Targets in melanoma. Receptor/ligand interactions in green are costimulatory and those in orange are coinhibitory. APC = Antigen-presenting cell.

melanoma patients, the median survival was 17 months, with an objective response rate of 31% [14]. A phase III trial (CheckMate 037) that randomized patients to chemotherapy versus nivolumab showed a higher response rate (32 vs. 10%) and duration of response (not reached vs. 3.5 months) [15] with nivolumab; a phase III study (CheckMate 066) of nivolumab versus dacarbazine showed a 40% response rate in the nivolumab group versus 13.9% with dacarbazine [16]. A phase I study of pembrolizumab including 135 patients with advanced melanoma showed a 38% response rate, without a significant difference between those patients who were ipilimumab pretreated versus naïve [17]. In September 2014, the FDA approved pembrolizumab at a dose of 2 mg/kg every 3 weeks for patients who progressed on ipilimumab noting that it should be used after progression on a BRAF inhibitor and ipilimumab in patients with BRAF-mutant melanoma. Anti-PD-L1 antibodies are under active investigation in metastatic melanoma, with promising results for MPDL3280A (Genentech) [18] and MEDI4736 (AstraZeneca) [19].

Reeducating T Cells

Another method to activate T cells against melanoma is to alter or augment their intrinsic behavior. Dr. Steven Rosenberg's group has pioneered a strategy of tumor resection and tumor-infiltrating lymphocyte (TIL) isolation, lymphodepletion, followed by TIL infusion then high-dose IL-2. Three small studies that included a total of 93 patients with metastatic melanoma who received TIL infusion yielded response rates ranging from 49 to 72% [20].

Intralesional therapy has a long history in melanoma, with talimogene laherparepvec (T-VEC) being the first agent to demonstrate activity in a phase III trial. This agent is an attenuated oncolytic herpes simplex virus encoding granulocyte macrophage colony-stimulating factor (GM-CSF) that is injected intratumorally, resulting in both a direct oncolytic effect and a systemic immune response. In a phase III trial of 436 patients with stage III or limited stage IV melanoma randomized to T-VEC versus subcutaneous GM-CSF injections, the overall response rate was 26.5 versus 5.7%, and overall survival was also increased in patients with stage III or IVA disease [21]. In a similarly designed phase II study, intralesional administration of CAVATAK, a Coxsackie A21 virus, was also demonstrated to result in responses at distant tumor sites [22].

The most promising data for vaccine therapy came from a randomized phase III trial of 185 patients with stage IV or locally advanced stage III melanoma treated with IL-2 with or without the gp100 vaccine plus Montanide-ISA51 adjuvant [23]. Patients treated with the vaccine plus IL-2 experienced an increase in median overall survival from 11.1 to 17.8 months ($p = 0.06$) when compared with IL-2 alone.

Combination Therapies

To achieve successful immune recognition and killing of melanoma, several combinatorial strategies have been evaluated in order to enhance antigen release and presentation, reverse T cell dysfunction or inhibit immune compensatory mechanisms. Conventional therapies, such as radiation, cytotoxic drugs and targeted agents, have been explored as a means to enhance antigen release and immune activation (table 1). While there is a concern that these approaches could be counterproductive by theoretically eliminating the immune cells, multiple studies have now demonstrated that in addition to direct cytotoxic effects on cancer cells, they may actually promote immunogenic cell death and the activation of antitumor immune response [24–28]. Several case reports have demonstrated that in patients treated with anti-CTLA-4 therapy, localized radiation therapy can lead to abscopal effects [29–32]. Both ipilimumab and nivolumab have been explored in combination with chemotherapy in several trials [33–36], though in some studies with increased toxicity [36]. Ipilimumab was also explored in combination with the VEGF inhibitor bevacizumab in a clinical trial of 46 patients in melanoma, demonstrating a notable 67% disease control rate [37]. There is also accumulating evidence that tyrosine kinase inhibitors may exert additional immunomodulatory effects. RAF inhibitors result in T cell activation and proliferation [38], whereas MEK inhibitors suppress the T cell receptor function and reduce antigen crosspresentation [39]. In patients, the combination of vemurafenib and ipilimumab resulted in hepatotoxicity, necessitating trial discontinuation [40] and severe cutaneous toxicities were seen in patients who received vemurafenib after nivolumab or ipilimumab therapy [41, 42]. Overall, these findings highlight the markedly

Table 1. Selected combination therapies with promising clinical data in metastatic melanoma

Agents	Reference
Anti-CTLA-4 + radiation	[29–32]
Ipilimumab or nivolumab + chemotherapy	[33–36]
Ipilimumab + nivolumab	[50, 51]
Ipilimumab + bevacizumab	[37]
Toll-like receptor agonists + type I interferon	[47]
Ipilimumab + intralesional talimogene laherparepvec	[49]
Ipilimumab + anti-IDO	[55]

complex nature of the effects of cytotoxic and targeted therapies when combined with immunotherapies.

In melanoma, as an additional strategy to enhance tumor antigen presentation, vaccination has been explored in combination with immune checkpoint blockades. Despite promising early results [43], in a phase III study, the addition of the gp100 peptide vaccine did not prove to be beneficial compared to ipilimumab alone [44]. As an alternative strategy to directly activating innate immunity, approaches using Toll-like receptor agonists and type I interferon have also been explored in combination with immunomodulatory antibodies with good tolerability and a suggestion of increased clinical benefit seen in all studies [45–47]. Several studies have explored oncolytic viruses as an in situ vaccination strategy to potentiate the efficacy of immunomodulatory antibody therapy. In the murine B16-F10 melanoma model, intratumoral administration of oncolytic Newcastle disease virus with systemic CTLA-4 blockade led to regression of distant tumor lesions and cures in several challenging tumor models [48]. As a clinical confirmation of these findings, the preliminary data from the phase I study of the combination of intralesional T-VEC with systemic ipilimumab demonstrated an objective response rate of 41% [49].

Several studies are exploring combination therapies of anti-CTLA-4 with other agents targeting T cell costimulatory or coinhibitory receptors. In a phase I clinical trial in patients with advanced melanoma where ipilimumab was evaluated in combination with nivolumab, over half of the patients experienced tumor reductions of over 80% [50], which were durable at both the 1- and 2-year follow-up [51]. Upregulation of the indoleamine 2,3-dioxygenase (IDO) in the tumor microenvironment is both a possible mechanism of resistance to anti-CTLA-4 immunotherapy [52] and a possible marker of response [53]. A combination of small-molecule inhibitors of IDO with CTLA-4-blocking antibodies resulted in significant enhancement of therapeutic efficacy in mouse models and patients [52, 54, 55]. Additional studies are focused on targeting other immunosuppressive populations in the tumor microenvironment, such as regulatory T cells and myeloid-derived suppressor cells, with preclinical studies demonstrating the synergy of such strategies with immune checkpoint blockade [56–58].

Conclusions and Future Directions

Combination therapy with pembrolizumab and now potentially with the combination of ipilimumab and nivolumab has raised the bar for clinical activity in melanoma [50]. Ongoing targets of preclinical and early-phase trials include OX40, 4-1BB, and GITR, LAG-3, ICOS, TIM-3, CD40 and the NK inhibitory receptor killer immunoglobulin-like receptors. The elegant and dynamic complexity of the interactions between tumors and the immune system presents both a challenge and an opportunity for the field of immunotherapy in melanoma.

References

1 Dutcher JP, et al: A phase II study of interleukin-2 and lymphokine-activated killer cells in patients with metastatic malignant melanoma. J Clin Oncol 1989;7:477–485.

2 Rosenberg SA, et al: Regression of established pulmonary metastases and subcutaneous tumor mediated by the systemic administration of high-dose recombinant interleukin 2. J Exp Med 1985;161: 1169–1188.

3 Rosenberg SA, et al: Treatment of 283 consecutive patients with metastatic melanoma or renal cell cancer using high-dose bolus interleukin 2. JAMA 1994; 271:907–913.

4 Creagan ET, et al: Phase II trials of recombinant leukocyte A interferon in disseminated malignant melanoma: results in 96 patients. Cancer Treat Rep 1986; 70:619–624.

5 Mocellin S, et al: Interferon alpha adjuvant therapy in patients with high-risk melanoma: a systematic review and meta-analysis. J Natl Cancer Inst 2010;102: 493–501.

6 Kirkwood JM, et al: A pooled analysis of Eastern Cooperative Oncology Group and Intergroup trials of adjuvant high-dose interferon for melanoma. Clin Cancer Res 2004;10:1670–1677.

7 Flaherty LE, et al: Southwest Oncology Group S0008: a phase III trial of high-dose interferon alfa-2b versus cisplatin, vinblastine, and dacarbazine, plus interleukin-2 and interferon in patients with high-risk melanoma – an Intergroup Study of Cancer and Leukemia Group B, Children's Oncology Group, Eastern Cooperative Oncology Group, and Southwest Oncology Group. J Clin Oncol 2014;32:3771–3778.

8 Hodi FS, et al: Improved survival with ipilimumab in patients with metastatic melanoma. N Engl J Med 2010;363:711–723.

9 Robert C, et al: Ipilimumab plus dacarbazine for previously untreated metastatic melanoma. N Engl J Med 2011;364:2517–2526.

10 Camacho LH, et al: Phase I/II trial of tremelimumab in patients with metastatic melanoma. J Clin Oncol 2009;27:1075–1081.

11 Kirkwood JM, et al: Phase II trial of tremelimumab (CP-675,206) in patients with advanced refractory or relapsed melanoma. Clin Cancer Res 2010;16:1042–1048.

12 Ribas A, et al: Phase III randomized clinical trial comparing tremelimumab with standard-of-care chemotherapy in patients with advanced melanoma. J Clin Oncol 2013;31:616–622.

13 Eggermont AM, Chiarion-Sileni V, Grob JJ, Dummer R, Wolchok JD, Schmidt H, Hamid O, Robert C, Ascierto PA, Richards JM, Lebbe C, Ferraresi V, Smylie M, Weber JS, Maio M, Konto C, Gurunath RK, de Pril V, Suciu S, Testori A: Ipilimumab versus placebo after complete resection of stage III melanoma: initial efficacy and safety results from the EORTC 18071 phase III trial (abstract). J Clin Oncol 2014;32(suppl 5):LBA9008.

14 Topalian SL, Sznol M, McDermott DF, et al: Survival, durable tumor remission, and long-term safety in patients with advanced melanoma receiving nivolumab. J Clin Oncol 2014;32:1020–1030.

15 Weber JS, Minor DR, D'Angelo SP, Hodi FS, Gutzmer R, et al: A phase 3 randomized, open-label study of nivolumab versus investigator's choice of chemotherapy in patients with advanced melanoma after prior anti-CTLA4 therapy. Ann Oncol 2014, DOI: 10.1093/annonc/mdu438.34.

16 Robert C, Long GV, Brady B, et al: Nivolumab in previously untreated melanoma without BRAF mutation. N Engl J Med 2015;372:320–330.

17 Hamid O, et al: Safety and tumor responses with lambrolizumab (anti-PD-1) in melanoma. N Engl J Med 2013;369:134–144.

18 Hamid O, Sosman JA, Lawrence DP, Sullivan RJ, Ibrahim N, Kluger HM, Boasberg PD, Flaherty K, Hwu P, Ballinger M, MOkatrin A, Kowanetz M, Chen DS, Hodi FS: Clinical activity, safety, and biomarkers of MPDL3280A, an engineered PD-L1 antibody in patients with locally advanced or metastatic melanoma (abstract). J Clin Oncol 2013;31(suppl): 9010.

19 Segal NH, Antonia SJ, Brahmer JR, Maio M, Blake-Haskins A, Li X, Rami JV, Ibrahim RA, Lutzky J, Khleif S: Preliminary data from a multi-arm expansion study of MEDI4736, an anti-PD-L1 antibody (abstract). J Clin Oncol 2014;32(suppl 5):3002.

20 Rosenberg SA, et al: Durable complete responses in heavily pretreated patients with metastatic melanoma using T-cell transfer immunotherapy. Clin Cancer Res 2011;17:4550–4557.

21 Kaufman HL, Andtbacka RHI, Collichio FA, Amatruda T, Senzer NN, Chesney J, delman KA, Spitler LE, Puzanov I, Ye Y, Li A, Gansert JL, Coffin R, Ross MI: Primary overall survival (OS) from OPTiM, a randomized phase III trial of talimogene laherparepvec (T-VEC) versus subcutaneous (SC) granulocyte-macrophage colony-stimulating factor (GM-CSF) for the treatment (tx) of unresected stage IIIB/C and IV melanoma (abstract). J Clin Oncol 2014;32(suppl 5):9008a.

22 Andtbacka RH, Curti BD, Kaufman H, Daniels G, Nemunaitis JJ, Spitler LE, Hallmeyer S, Lutzky J, Schultz S, Whitman ED, Zhou K, Weisberg JI, Schafren D: CALM study: a phase II study of an intratumorally delivered oncolytic immunotherapeutic agent, coxsackievirus A21, in patients with stage IIIc and stage IV malignant melanoma (abstract). J Clin Oncol 2014;32(suppl 5):3031.

23 Schwartzentruber DJ, et al: gp100 peptide vaccine and interleukin-2 in patients with advanced melanoma. N Engl J Med 2011;364:2119–2127.

24 Zitvogel L, Kepp O, Kroemer G: Immune parameters affecting the efficacy of chemotherapeutic regimens. Nat Rev Clin Oncol 2011;8:151–160.

25 Tesniere A, et al: Immunogenic cancer cell death: a key-lock paradigm. Curr Opin Immunol 2008;20: 504–511.

26 Kroemer G, Zitvogel L: Abscopal but desirable: the contribution of immune responses to the efficacy of radiotherapy. Oncoimmunology 2012;1:407–408.

27 Gameiro SR, et al: Radiation-induced immunogenic modulation of tumor enhances antigen processing and calreticulin exposure, resulting in enhanced T-cell killing. Oncotarget 2014;5:403–416.

28 Obeid M, et al: Calreticulin exposure is required for the immunogenicity of γ-irradiation and UVC light-induced apoptosis. Cell Death Differ 2007;14:1848–1850.

29 Postow MA, et al: Immunologic correlates of the abscopal effect in a patient with melanoma. New Engl J Med 2012;366:925–931.

30 Golden EB, et al: An abscopal response to radiation and ipilimumab in a patient with metastatic non-small cell lung cancer. Cancer Immunol Res 2013;1: 365–372.

31 Hiniker SM, et al: A systemic complete response of metastatic melanoma to local radiation and immunotherapy. Transl Oncol 2012;5:404–407.

32 Sullivan RJ, et al: Case records of the Massachusetts General Hospital: case 21-2013. A 68-year-old man with metastatic melanoma. New Engl J Med 2013; 369:173–183.

33 Reck M, et al: Ipilimumab in combination with paclitaxel and carboplatin as first-line therapy in extensive-disease-small-cell lung cancer: results from a randomized, double-blind, multicenter phase 2 trial. Ann Oncol 2013;24:75–83.

34 Robert C, et al: Ipilimumab plus dacarbazine for previously untreated metastatic melanoma. New Engl J Med 2011;364:2517–2526.

35 Lynch TJ, et al: Ipilimumab in combination with paclitaxel and carboplatin as first-line treatment in stage IIIB/IV non-small-cell lung cancer: results from a randomized, double-blind, multicenter phase II study. J Clin Oncol 2012;30:2046–2054.

36 Antonia SJ, et al: Nivolumab (anti-PD-1; BMS-936558, ONO-4538) in combination with platinum-based doublet chemotherapy (PT-DC) in advanced non-small cell lung cancer (NSCLC) [abstract]. J Clin Oncol 2014;32(suppl 5):8113.

37 Hodi FS, et al: Bevacizumab plus ipilimumab in patients with metastatic melanoma. Cancer Immunol Res 2014;2:632–642.

38 Callahan MK, et al: Paradoxical activation of T cells via augmented ERK signaling mediated by a RAF inhibitor. Cancer Immunol Res 2014;2:70–79.

39 Vella LJ, et al: MEK inhibition, alone or in combination with BRAF inhibition, affects multiple functions of isolated normal human lymphocytes and dendritic cells. Cancer Immunol Res 2014;2:351–360.

40 Ribas A, et al: Hepatotoxicity with combination of vemurafenib and ipilimumab. New Engl J Med 2013; 368:1365–1366.

41 Johnson DB, et al: Severe cutaneous and neurologic toxicity in melanoma patients during vemurafenib administration following anti-PD-1 therapy. Cancer Immunol Res 2013;1:373–377.

42 Harding JJ, Pulitzer M, Chapman PB: Vemurafenib sensitivity skin reaction after ipilimumab. New Engl J Med 2012;366:866–868.

43 Sanderson K, et al: Autoimmunity in a phase I trial of a fully human anti-cytotoxic T-lymphocyte antigen-4 monoclonal antibody with multiple melanoma peptides and Montanide ISA 51 for patients with resected stages III and IV melanoma. J Clin Oncol 2005;23:741–750.
44 Hodi FS, et al: Improved survival with ipilimumab in patients with metastatic melanoma. New Engl J Med 2010;363:711–723.
45 Millward M, et al: Phase 1 study of tremelimumab (CP-675 206) plus PF-3512676 (CPG 7909) in patients with melanoma or advanced solid tumours. Br J Cancer 2013;108:1998–2004.
46 Kudchadkar RR, et al: A phase IB study of ipilimumab with peginterferon alfa-2b in patients with unresectable melanoma (abstract). J Clin Oncol 2013; 31(suppl):9079.
47 Tarhini AA, et al: Safety and efficacy of combination immunotherapy with interferon alfa-2b and tremelimumab in patients with stage IV melanoma. J Clin Oncol 2012;30:322–328.
48 Zamarin D, et al: Localized oncolytic virotherapy overcomes systemic tumor resistance to immune checkpoint blockade immunotherapy. Sci Transl Med 2014;6:226ra32.
49 Puzanov I, et al: Primary analysis of a phase 1b multicenter trial to evaluate safety and efficacy of talimogene laherparepvec (T-VEC) and ipilimumab (ipi) in previously untreated, unresected stage IIIB-IV melanoma (abstract). J Clin Oncol 2014; 32(suppl 5):9029.
50 Wolchok JD, et al: Nivolumab plus ipilimumab in advanced melanoma. N Engl J Med 2013;369:122–133.
51 Sznol M, et al: Survival, response duration, and activity by BRAF mutation (MT) status of nivolumab (NIVO, anti-PD-1, BMS-936558, ONO-4538) and ipilimumab (IPI) concurrent therapy in advanced melanoma (MEL) [abstract]. J Clin Oncol 2014; 32(suppl 5):LBA9003.
52 Holmgaard RB, et al: Indoleamine 2,3-dioxygenase is a critical resistance mechanism in antitumor T cell immunotherapy targeting CTLA-4. J Exp Med 2013; 210:1389–1402.
53 Hamid O, et al: A prospective phase II trial exploring the association between tumor microenvironment biomarkers and clinical activity of ipilimumab in advanced melanoma. J Transl Med 2011;9:204.
54 Spranger S, et al: Mechanism of tumor rejection with doublets of CTLA-4, PD-1/PD-L1, or IDO blockade involves restored IL-2 production and proliferation of CD8+ T cells directly within the tumor microenvironment. J Immunother Cancer 2014;2:3.
55 Gibney GT, et al: Preliminary results from a phase 1/2 study of INCB024360 combined with ipilimumab (ipi) in patients (pts) with melanoma (abstract). J Clin Oncol 2014;32(suppl 5):3010.
56 Sutmuller RP, et al: Synergism of cytotoxic T lymphocyte-associated antigen 4 blockade and depletion of $CD25^+$ regulatory T cells in antitumor therapy reveals alternative pathways for suppression of autoreactive cytotoxic T lymphocyte responses. J Exp Med 2001;194:823–832.
57 Kim K, et al: Eradication of metastatic mouse cancers resistant to immune checkpoint blockade by suppression of myeloid-derived cells. Proc Natl Acad Sci U S A 2014;111:11774–11779.
58 Zhu Y, et al: CSF1/CSF1R blockade reprograms tumor-infiltrating macrophages and improves response to T cell checkpoint immunotherapy in pancreatic cancer models. Cancer Res 2014;74:5057–5069.

Jedd D. Wolchok, MD
Memorial Sloan-Kettering Cancer Center
Ludwig Center for Cancer Immunotherapy
1275 York Avenue, New York, NY 10065 (USA)
E-Mail wolchokj@mskcc.org

Michielin O, Coukos G (eds): Immuno-Oncology.
Prog Tumor Res. Basel, Karger, 2015, vol 42, pp 30–43 (DOI: 10.1159/000437183)

Immunotherapy of Breast Cancer

Carmen Criscitiello · Giuseppe Curigliano
Experimental Therapeutics Division, European Institute of Oncology, Milan, Italy

Abstract
Cancer immunoediting is the process by which the immune system protects the host from tumor development and guides the somatic evolution of tumors by eliminating highly immunogenic tumor cells. A fundamental dogma of tumor immunology and of cancer immunosurveillance in particular is that cancer cells express antigens that differentiate them from their nontransformed counterparts. Molecular studies clearly show that these antigens were often products of mutated cellular genes, aberrantly expressed normal genes, or genes encoding viral proteins. There is a strict correlation between genetic instability and the immune landscape of a breast cancer. Mutational heterogeneity in breast cancer is associated with new cancer-associated genes and new cancer antigens. Frequencies of somatic mutations or mutational burden can be related to the immunogenicity of breast cancer. We believe that molecular subtypes of breast cancer that are triple negative, luminal B-like or HER2-positive have a high mutational burden and can be considered immunogenic. The increasing knowledge of the immune system's capacity to not only recognize and destroy cancer, but also to shape cancer immunogenicity will develop more informed attempts to control cancer via immunological approaches. To be effective in breast cancer, immunotherapies will have to increase the quality or quantity of immune effector cells, reveal additional protective tumor antigens, and/or eliminate cancer-induced immunosuppressive mechanisms. Multiple immunotherapy approaches are under investigation in patients with breast cancer. These include vaccine approaches to elicit strong specific immune responses to tumor antigens such as WT-1, HER2 and NY-ESO-1, approaches involving adoptive transfer of in vitro-expanded, naturally arising or genetically engineered tumor-specific lymphocytes, therapeutic administration of monoclonal antibodies to target and eliminate tumor cells, and approaches that inhibit or destroy the molecular or cellular mediators of cancer-induced immunosuppression, such as CTLA-4, PD-1 or Treg cells. Here we provide a concise and comprehensive review on the role and utility of promising immunotherapeutics for the treatment of patients with breast cancer.

Immunotherapy in breast cancer (BC) primarily aims at allowing the immune system to identify neoplastic growth and to prevent carcinogenesis and eliminate transformed cells which could potentially become malignant tumor cells. Both chemotherapy and targeted agents can influence the immune system. Immune response may not only lead to the rejection of cancer, but can also exert regulatory effects promoting tumor growth. According to the cancer immunoediting hypothesis, the immune system not only protects the host against tumor formation, but also shapes tumor immunogenicity [1]. The immunoediting hypothesis stresses the dual host-protective and tumor-promoting actions of immunity on developing tumors [1]. This concept of immunoediting in cancer immunology has a profound impact on how we understand some of the long-term efficacy of chemotherapy and radiotherapy, treatments formerly believed to be highly immunosuppressive. The combined understanding of tumor and host response may very likely lead to extended immunotherapeutic advances in BC. Despite immune surveillance, solid tumors are frequently dimly immunogenic and do appear to circumvent detection by the various effector cells of the immune system or have been able to limit the extent of immunological surveillance, thereby evading eradication [2]. A number of mechanisms are involved in immune evasion and the equilibrium between immune and autoimmune responses requires further investigation. More composite and sophisticated mechanisms operate through the recruitment of inflammatory cells that are actively immunosuppressive, including regulatory T cells (Tregs). They can suppress cytotoxic lymphocytes; therefore, cancer may develop as a result of selection of less immunogenic tumor cells (immunoediting) and/or due to augmented effectiveness of tumor-mediated immunosuppression (immune subversion) [1, 3]. There is a dynamic interplay between immune cells and tumor progression, which is currently under investigation. Accumulated data indicate that the success of an immune response in cancer can be determined by the type of immune response elicited. In BC, immune-related factors are crucial in defining both patient prognosis and response to treatment. These include, among others, the extent of lymphocytic infiltration in BC tissue and a class of gene expression signatures, both having the potential to more precisely characterize a patient's natural history and identify subgroups of patients with different sensitivities to given treatments. Despite the increasing knowledge in the field of immunology applied to cancer, medical oncologists still mostly rely on classical clinical-pathological features, such as tumor size and lymph node involvement, for daily patient management, and it is still difficult to foresee how these parameters might be implemented in daily clinical practice in the near future. Our aim is to provide the reader with a concise yet comprehensive review on the role and utility of promising immunotherapeutics for the treatment of patients with BC.

Lymphocytic Infiltrate in Breast Cancer

Over the past few decades, a growing body of evidence has emerged demonstrating that the immune system participates both in tumor development (via chronic inflammation orchestrated by the innate immune system) and in tumor elimination and control (through the actions of the adaptive immune system) [4]. The presence of tumor-infiltrating lymphocytes (TILs) is observed in some breast tumors and has been reported to be a good prognosis feature for some forms of the disease [5–7], particularly for rapidly proliferating tumors correlating with a negative axillary nodal status, smaller tumor size and a lower grade [6]. Similarly, the TIL count has been associated with better survival in patients with ER-negative tumors [8, 9]. Also, TILs have been negatively correlated with patient age at diagnosis [8, 9]. Recently, TILs have emerged as a potential prognostic and predictive marker in BC, especially in the triple-negative (TN) and HER2-positive subtypes. Loi et al. [10] have evaluated the predictive value of TIL in 935 patients within the FinHER trial. Among the 134 TN patients receiving docetaxel and FEC or vinorelbine and FEC, the 3-year recurrence-free survival was 90% in the case of extensive lymphocyte infiltrate (intratumoral or stromal TIL >30%) versus 66% in the case of nonextensive lymphocyte infiltrate ($p = 0.007$). In locally advanced BC treated with neoadjuvant chemotherapy, the presence of TILs in the primary biopsy predicts a pathologic complete response [9]. In this study, TIL was composed of both CD3+ and CD20+ cells. The same group has evaluated the prognostic and predictive value of TILs in a large cohort of lymph node-positive early BC patients prospectively randomized to receive either high-dose anthracycline-based chemotherapy or a combinatorial regimen involving anthracyclines and docetaxel within the BIG02-98 trial [11]. TILs+ were strongly associated with a good prognosis among TNBC patients, while TILs had no prognostic value among HER2-positive BC patients. However, TILs+ were associated with improved therapeutic responses to high-dose anthracyclines in patients with HER2-positive BC [11]. More recently, Adams et al. [12] reported on the prognostic value of TILs evaluated on samples from 481 patients with TNBC enrolled onto two prospective Eastern Cooperative Oncology Group (ECOG) adjuvant clinical trials (E2197 and E1199). This prospective study replicated findings from previous reports, thus confirming that TILs are positively associated with prognosis in primary TNBC. Also, Demaria et al. [13] reported that taxane-based primary chemotherapy converted 7 out of 21 breast tumors from TIL– to TIL+, and the postchemotherapy TIL+ status was associated with an improved clinical response. Furthermore, Ladoire et al. [14] reported that neoadjuvant chemotherapy increased the CD8+ infiltration in the tumor. This post-treatment infiltration is associated with an improved outcome. Moreover, Dieci et al. [15] showed that the presence of TIL in residual disease after neoadjuvant chemotherapy is associated with a better prognosis in TNBC patients. Interestingly, 23 out of 27 patients (85%) who were TIL+ after neoadjuvant chemotherapy were TIL– at baseline [15]. It is worth highlighting that one of the proposed mechanisms of action of chemotherapy is the induction of an antitumoral immune response. Zitvogel and colleagues [16] have reported that cytotoxic agents, including anthracyclines, oxaliplatin and radiation therapy, induce immunogenic cell death through high-mobility group box-1 (HMGB1) release. In this model chemotherapy kills cancer cells. This modality of cell death itself induces signal danger and tumor-specific T cell response. Interestingly, the induction of T cell response is highly heterogeneous according to individual and tumor characteristics. This interindividual variability in the ability of chemotherapy to induce immune response could explain why chemotherapy is not working the same in all patients. Overall, these findings provide evidence [14] that TIL assessed at baseline could stratify patients into high- or low-risk populations. Moreover, small retrospective hypothesis-generating studies suggested that post-chemotherapy lymphocyte infiltration at the tumor bed may be correlated with prognosis. In summary, TIL+ BCs present specific features that could have specific clinical implications in terms of prognosis. These data from population-based studies or randomized trials will need additional validation using a prospective trial before implementation. A potential implication could be to use TILs as inclusion criteria in large adjuvant trials that include TNBC or HER2-positive BC. Another potential prognostic implication relates to the potentially good prognosis of patients who were TIL– at baseline, but became TIL+ after neoadjuvant therapy [13]. Nevertheless, if these data are confirmed, they could be used to better stratify which patients should be included in postneoadjuvant trials. Finally, the information that TIL+ is associated with a good outcome is an argument to develop immune strategies in patients with BC. TIL assessment could identify a subset of TIL– TN and HER2-positive BC patients who require additional therapy. A genomic predictor of chemoimmunization could represent a new prognostic parameter that will allow selecting which TIL– BCs are candidates for new therapeutics in the (neo)adjuvant setting, thus decreasing toxicity, decreasing cost and making feasible (neo)adjuvant trials. TILs are mainly represented by nonactivated T cells which often become activated after exposure to chemotherapy. More recently, the nature of TILs has been better characterized. Ruffell et al. [17] reported that TILs are mainly comprised of CD3+/CD56– T cells, while a minority consisted in natural killer (NK) cells or CD20+ cells. The majority of CD3+ cells were either CD4+ or CD8+ T cells. Interestingly, CD8+ cells did not express granzyme B at baseline, which means that they did present inactivation status, while they changed to express granzyme B after neoadjuvant chemotherapy in one third of the patients. Lastly, a minority of TILs presented T and NK cell features. The genomic characteristics of TIL+ tumors are important to understand which molecular mechanisms lead to lymphocyte infiltration. Genomic instability may promote an antitumor immune response through tumor-associated antigens (TAAs). Some mechanisms of chemokine release by the tumor have been described and correlated with lymphocyte attraction. TILs have been associated with CXCL9 and CXCL13 expression by the tumor [9]. TIL+ tumors present a specific methylation pattern on immune-related genes, including CCL5 [18], and a cluster of chemokines is lost in a subset of BC [19].

What are the clinical implications of all 'TILs' data produced in the last years? First, validation of whether TILs are prognostic or predictive in HER2-positive BC is needed, preferably in a large population set with an appropriate follow-up time. Second, there is a need to validate immune genomic signatures that may be predictive and prognostic in patients with TN and HER2-positive disease. Third, we support the concept of developing and potentially incorporating an 'immunoscore' into traditional classifications of BC, thus providing an essential prognostic and potentially predictive tool in the pathology report. Fourth, clinical trials for TN and HER2-positive BC in the metastatic setting with drugs that target immune-cell-intrinsic checkpoints should be implemented. Blockade of one of these checkpoints, cytotoxic T lymphocyte-associated antigen 4 (CTLA-4) or the programmed death-1 (PD-1) receptor may provide proof of concepts for the activity of an immune-modulation approach in the treatment of a BC.

Immune-Related Gene Signatures in Breast Cancer

In recent years, gene expression profiling has been used in an effort to more precisely define BC taxonomy and identify prognostic and predictive signatures [20]. The common denominator between the majority of the 'first generation' signatures is their overall capacity to detect subtle differences in the cell cycle and proliferation. For this reason, they have not been found to be prognostic in the TN or HER2-positive subtypes since these tumors are by 'nature' highly proliferative. Several investigators have tried to overcome the limitations of these 'first generation' signatures by focusing on the BC microenvironment and/or immune response to define promising 'second generation' prognostic signatures (table 1) [21]. Unsupervised gene expression profiling of cancer-associated stroma revealed a signature enriched for CD8+ T cell responses that was predictive of a good prognosis [22]. The STAT1 immune response module has been shown to associate with survival in TN and HER2-positive patients [23, 24], and in the same BC subtypes the overexpression of immune-related genes was able to identify subgroups of patients with a better prognosis [7, 25]. Together, these investigations suggest that immune modulation may be important particularly in highly proliferative subtypes such as TN and HER2-positive.

Breast Cancer Chemotherapy and the Antitumor Immune Response

The role of the immune system extends beyond its direct interactions with tumor cells and its battle with tumor progression. Numerous recent reports have highlighted the complex interactions that occur between the immune response and cytotoxic drugs used to treat cancer patients. Both players have the ability to potentially enhance or harm their counterpart and thereby detrimentally affect their antitumor functions [26, 27]. There is accumulating evidence that some cytotoxic drugs, such as anthracyclines, actually promote antitumor immunity and thereby contribute to the treatment's therapeutic effect. This is mainly linked with the ability of these drugs to provoke 'immunogenic' cell death, with the apoptotic cancer cells emitting signals that trigger antitumor immunity [28–30]. In response to the insults induced by cytotoxic agents, tumor cells translocate intracellular calreticulin to the cell surface where it

Table 1. Immune signatures and their development

Immune signature	Signature development
Immune response (IR) module [25]	A subclass of estrogen receptor (ER)-negative tumors that overexpress IR genes and that have a good prognosis compared with other ER-negative breast tumors independent of lymph node status or lymphocytic infiltration was identified. Subsequently, an associated module of complement and IR genes that define prognostic markers was identified and validated in over 210 ER negative samples
STAT1 module [24]	Based on the literature, genes to act as 'prototypes' for different biological processes (ER for ER signaling, HER2 for HER2 signaling, AURKA for proliferation, CASP3 for apoptosis, VEGF for angiogenesis, PLAU for tumor invasion/metastasis and, in this case, STAT1 for IR) were selected. A comparison of linear models was then applied to generate modules of genes specifically associated with each of the prototype genes but not with the other prototypes
B cell metagene [7]	Gene expression patterns of 200 patients who did not receive systemic treatment and coregulated genes related to proliferation, steroid hormone receptor expression and B and T cell infiltration were identified after performing hierarchical cluster analysis. Metagenes were calculated as a surrogate for all genes contained within a particular cluster and their expression was correlated to time-to-metastasis. The B cell metagene showed independent prognostic information in carcinoma with high proliferative activity
IgG, HCK, MHC-I, MHC-II, LCK, STAT1, interferon metagenes [78]	Unsupervised hierarchical clustering of genes in 12 primary invasive BC datasets as well as combined datasets revealed a large cluster of genes with functions in immune cells. Among it, clusters that contained a minimum number of elements and a minimal average correlation were selected and 7 metagenes were derived. Each metagene was then associated with a cell type and/or immunological state
HR-negative/T-negative signature [79]	A cohort of node-negative, adjuvant treatment-naïve hormone receptor-negative (HR-negative) and TN (T-negative) BC patients has been used to define and validate genes predictive for distant metastatic relapse. A composite HR-negative/T-negative signature index was able to identify cases likely to remain free of metastatic relapse with high accuracy. Of note, significant positive correlation was observed between the HR-negative/T-negative index and three independent immune-related signatures (STAT1, IFN, IR), and network analysis showed that the signature was linked to immune/inflammatory cytokine regulation
SVM classifier [80]	Gene expression data of 2,145 invasive early breast adenocarcinomas were collected and used to test and validate the predictive performance of a support vector machine (SVM) classifier based on a 368-gene expression signature associated with medullary breast carcinoma (MBC), which displays a basal profile but has good prognosis. The SVM model accurately classified all MBC samples in the learning and validation sets and was able to separate 466 cases of basal BCs into two subgroups, subgroup 1 and subgroup 2, containing good- and poor-prognosis tumors, respectively. Ontology analysis revealed, among other features, effective IR in the good-prognosis subgroup

functions as an 'eat me' signal for dendritic cells (DC), thereby provoking immunogenic cell death. Molecular chaperones such as the heat-shock protein 90 can also be induced, appearing on the tumor cell surface where they contribute to adhesion with DC, a first step in DC maturation. Apoptotic tumor cells also release HMGB1 (chromatin-binding protein HMGB1), which mediates Toll-like receptor 4-dependent processing of the DC phagocytic cargo and the release of maturation signals, including type I interferons or interleukin-1 necessary for the completion of DC maturation [16, 31–33]. Doxorubicin was also shown to increase tumor antigen-specific proliferation of CD8 T cells, and in a population of BC patients who underwent biopsy before neoadjuvant anthracycline-based therapy a correlation between CD8α, CD8β and IFN-γ gene expression levels and clinical response was identified, underlying the importance of CD8 T cells and IFN-γ in defining the therapeutic effect of doxorubicin [34]. In contrast, drugs such as cisplatin and alkylating agents lead to 'nonimmunogenic' cell death, which does not elicit an immune response and is even considered tolerogenic [27, 35–37]. Many chemotherapeutic agents currently used in cancer treatment have immunosuppressive side effects. For example, cyclophosphamide and methotrexate have been shown to impair T cell proliferative and/or effector functions in the periphery [38, 39]. Glucocorticoids, used in standard practice to relieve chemotherapy-associated nausea and vomiting, have potent immunosuppressive properties that can repress DC differentiation and antigen presentation [40], NK cell activity and T cell development and function [41, 42]. The complexity increases when considering that cyclophosphamide used at low doses has an immunostimulatory effect by decreasing the number and inhibitory function of CD4+CD25+ Treg cells [43]. A daily low dose of oral cyclophosphamide (i.e. a metronomic schedule) when given to patients with end-stage cancer for 1 month suppressed Treg inhibitory functions and restored the proliferative capacity of effector T cells and NK cell cytotoxicity [44]. There is also some evidence that taxanes, usually considered immunosuppressive [45], can also stimulate the antitumor immune response. Using T cell proliferation and NK cell cytolysis assays, it was found that a cohort of patients with stage II/III BC treated with taxanes had enhanced T cell and NK cell function compared with patients whose treatment did not include taxanes [46]. A clinical response to neoadjuvant paclitaxel was associated with increased recruitment of TILs in malignant cell nests from patients with locally advanced BC [26]. Experimental murine models have shown that paclitaxel can also increase the efficacy of immunostimulatory agents [47, 48]. Also, antimetabolites used in the treatment of BC [e.g. 5-fluorouracil (5-FU), gemcitabine, methotrexate, etc.] have been shown to actively interact with the immune system, producing a variety of immunostimulatory effects [49–51]. 5-FU treatment of colon and breast carcinoma cells in vitro was shown to enhance the expression of the carcinoembryonic antigen (CEA) and the susceptibility to HLA-A(*)02.01-restricted, CEA-peptide-specific cytotoxic T cells. This restoration of tumor antigen expression provides a rationale for investigating a possible new role for 5-FU as an immune amplifying agent in breast and colorectal cancer patients immunized with CEA-directed cancer vaccines [52]. Alternatively, the immune system can directly influence the response to chemotherapy agents in many different ways. DCs express surface receptors, such as Toll-like receptor 4 and P2RX7, which specifically recognize soluble molecules released by apoptotic tumor cells (HMGB1 and adenosine triphosphate, respectively) and thereby play a key role directing their interactions with immune effectors. Studies have shown that when these receptors are either absent [33, 53] or affected by a loss-of-function single-nucleotide polymorphism [33, 54] the ability of the immune system to mount an antitumor response after chemotherapy is impaired in node-positive BC patients treated with anthracyclines.

Interactions between the Immune System and Targeted Therapies

Targeted therapies are increasingly being developed and used for the treatment of a wide variety of cancers. These drugs include monoclonal antibodies (mAb) and tyrosine kinase inhibitors, which have been shown to increase patient survival. In BC, the humanized immunoglobulin (IgG1) monoclonal antibody trastuzumab is widely used to treat patients with HER2-positive tumors and is able to increase the response rate and survival for this particular subtype of BC [55, 56]. Antibody-dependent cellular cytotoxicity (ADCC) has long been implicated as one of the mechanisms of action for trastuzumab [57, 58]. Correlative studies have suggested that patients responding to mAb treatment had higher in situ infiltration of leukocytes and increased capacity to mediate ADCC activity. Recent data also suggest that tumor regressions in vivo due to anti-HER2 mAb therapy also require an effective adaptive antitumor immune response in order to achieve optimal therapeutic effects; levels of CD8 and IFN-γ have been shown to correlate with HER2 targeted treatment. Preclinical models have shown that inhibitory signals mediated by coreceptors on tumor-specific T cells impede antitumor immunity and suggest that blockade of such activity may facilitate host immune-mediated tumor elimination. Therefore, immunotherapeutics that augment CD8 T cell antitumor activity, such as anti-PD-1 and anti-CTLA4 mAb, may improve outcomes in combination with trastuzumab in HER2 overexpressing BC patients by involving and enhancing critical host immunity [59–61]. PD-1 is an inhibitory receptor expression on activated T and B cells whose activity may suppress antitumor immunity. Phase I data has determined that blocking the PD-1 immune checkpoint with a targeted antibody has been shown to be safe and well tolerated, with evidence of antitumor activity in the phase I setting, presumably through immunogenic mechanisms as PD-1 receptors are not expressed on nonhematologic malignancies [62]. The combinatorial antitumor effects of anti-HER2 and anti-PD-1 have been demonstrated in experimental models and are important proof of principle experiments that the antitumor adaptive immunity provoked by anti-HER2 can be capitalized upon, providing a new paradigm of potential combination treatment for women receiving trastuzumab. Therefore, a combination with other immunotherapeutic approaches such as trastuzumab – which also acts through direct tumoricidal mechanisms – has been shown to be highly synergistic in mouse models and therefore warrants evaluation in human patients. In selected patients, such an approach may even eliminate the need for combination cytotoxic chemotherapy. CTLA-4 (CD152) bears homology to PD-1, but its immunoinhibitory signals are different. CTLA-4 knockout mice display early lethality, unlike PD-1 knockouts which demonstrate late-onset and organ-specific autoimmunity. Anti-CTLA4 mAb treatment has been extensively evaluated in humans and has resulted in durable tumor responses in phase III settings. However, in a phase III trial in metastatic melanoma, a significant frequency of adverse events was observed [63]. Similarly, however, combination anti-CTLA-4 mAb with trastuzumab has shown synergy in preclinical mouse models [64]. Treatments using mAb such as trastuzumab are examples of the oncogene addition, which refers to the process of tumor cell death that can occur after inactivation of a single oncogene. This phenomenon provides the rationale for molecularly targeted therapies. However, complete tumor response after molecular targeted therapies requires a functioning immune system, pointing the way toward radically new combination therapies [61]. Data from human BC patients suggests that baseline immune responses may be important for determining those patients that could mount an appropriate adaptive immune response. This seems particularly important in the BC subgroup that overexpresses HER2. Quantification of baseline TILs in a given BC may therefore be able to identify a group of women most likely to benefit from immunotherapeutic approaches and deserves further investigation in this setting.

Vaccine-Based Therapies for Breast Cancer

Vaccines constitute an active and specific immunotherapy designed to stimulate the intrinsic antitumor immune response by presenting TAAs expressed on normal tissues that are overexpressed on tumor cells. Malignant cells can express both normal self-antigens and specific TAAs that arise from genetic mutations and/or epigenetic changes, recognized by the immune response either through their loss or de novo aberrant expression. Many TAAs (including MUC1, HER-2, CEA, hTER, WT1) have been identified and been shown to be specifically recognized by T cells [65]. Induction of strong immunity by cancer vaccines is expected to lead to the establishment of immunological memory, thereby preventing tumor recurrence.

The identification of immunological and genetic features affecting immune response in patients with a minimal tumor burden represents the optimal background for the development of clinical studies in the adjuvant setting as it should allow the immune system to mount a response before it is overtaken. An active immunization may provide a nontoxic therapeutic modality able to induce antitumor immune responses in cancer patients [66]. However, the majority of the trials so far have been conducted in the metastatic setting, which could have significantly impacted on the results due to the large tumor burden [67–71]. Cancer vaccines are more effective when given in combination with standard cancer treatments, which appear to increase their effectiveness [72–74]. The elimination of Tregs potentially provides the basis [70] for the synergistic effect between cancer vaccines and chemotherapy [73]. In order to optimize the immunological response to a vaccination strategy, it is paramount to identify both the target antigen and the patient population to be targeted. Large population analyses on specific BC subtypes are necessary to select patients who have a higher probability of expressing that specific antigen. In order to improve immunotherapy trials, the ability to initiate tumor-specific immunity, either directly by providing tumor-associated antigens or indirectly by favoring the cross-presentation of endogenous tumor antigens, the capacity to recruit effector immune cells within the tumor site by increasing tumor visibility, and the ability to preserve immune cell functionality within the tumor microenvironment through the subversion of immune-escape mechanisms must all be taken into account. In the adjuvant setting we can propose several surrogate markers of activity for a vaccination strategy. Some are related to immunomonitoring approaches, some others can be related to the detection of disseminated tumor cells in the bone marrow and/or circulating tumor cells in the peripheral blood [75, 76]. Combined therapies should be proposed to accomplish these three features. Furthermore, immunotherapy needs a considerably long time before turning into an efficient antitumor response, which is why trials should ideally target well-defined patients in adjuvant settings [77].

Conclusion

We have provided an overview of the recent evidence demonstrating that the immune system plays a critical role in defining BC prognosis and response to different treatments. However, more studies are necessary to precisely define the nature and role of TILs, establish the clinical value of immune-based prognostic and predictive gene signatures, and optimize standard treatments to effectively work in tandem with the antitumor immune response. The success of future immunotherapy strategies will depend on the identification of additional immunogenic antigens that can serve as the best tumor-rejection targets. It is also important to target antigens that are biologically important to tumor progression. Therapeutic success will also depend on developing the best antigen delivery systems and on the elucidation of the entire network of immune signaling pathways that regulate immune responses in the tumor microenvironment. Future challenges are also related to the identification of critical hurdles that can significantly delay clinical translation of advances in immunotherapy in cancer patient care. Some of these critical hurdles include: (a) the complexity of the cancer, tumor heterogeneity and immune escape; (b) lack of definitive biomarker(s) for assessment of clinical efficacy of cancer immunotherapies; (c) definition of new conventional clinical response criteria that take into consideration differences between response patterns to cytotoxic agents and immunotherapies, and (d) insufficient exchange of information critical to advancing the field. Addressing these hurdles will facilitate and improve the translation of novel immunotherapies to patients with cancer. A better understanding of the dialogue between cancer and immune cells could help to improve immunotherapeutic approaches and foster the development of new drugs that increase the ability of the immune response to provide effective and sustained antitumor immunity.

References

1 Schreiber RD, Old LJ, Smyth MJ: Cancer immunoediting: integrating immunity's roles in cancer suppression and promotion. Science 2011;331: 1565–1570.

2 Disis ML: Immune regulation of cancer. J Clin Oncol 2010;28:4531–4538.

3 Dunn GP, Old LJ, Schreiber RD: The three Es of cancer immunoediting. Ann Rev Immunol 2004;22: 329–360.

4 Zitvogel L, Tesniere A, Kroemer G: Cancer despite immunosurveillance: immunoselection and immunosubversion. Nat Rev Immunol 2006;6:715–727.

5 DeNardo DG, Coussens LM: Inflammation and breast cancer – balancing immune response: crosstalk between adaptive and innate immune cells during breast cancer progression. Breast Cancer Res 2007;9:212.

6 Aaltomaa S, Lipponen P, Eskelinen M, Kosma VM, Marin S, Alhava E, Syrjanen K: Lymphocyte infiltrates as a prognostic variable in female breast cancer. Eur J Cancer 1992;28A:859–864.

7 Schmidt M, Bohm D, von Torne C, Steiner E, Puhl A, Pilch H, Lehr HA, Hengstler JG, Kolbl H, Gehrmann M: The humoral immune system has a key prognostic impact in node-negative breast cancer. Cancer Res 2008;68:5405–5413.

8 Mahmoud SM, Paish EC, Powe DG, Macmillan RD, Grainge MJ, Lee AH, Ellis IO, Green AR: Tumor-infiltrating CD8+ lymphocytes predict clinical outcome in breast cancer. J Clin Oncol 2011;29:1949–1955.

9 Denkert C, Loibl S, Noske A, Roller M, Muller BM, Komor M, Budczies J, Darb-Esfahani S, Kronenwett R, Hanusch C, et al: Tumor-associated lymphocytes as an independent predictor of response to neoadjuvant chemotherapy in breast cancer. J Clin Oncol 2010;28:105–113.

10 Loi S, Michiels S, Salgado R, Sirtaine N, Jose V, Fumagalli D, Kellokumpu-Lehtinen PL, Bono P, Kataja V, Desmedt C, et al: Tumor infiltrating lymphocytes are prognostic in triple negative breast cancer and predictive for trastuzumab benefit in early breast cancer: results from the FinHER trial. Ann Oncol 2014;25:1544–1550.

11 Loi S, Sirtaine N, Piette F, Salgado R, Viale G, van Eenoo F, Rouas G, Francis P, Crown JP, Hitre E, et al: Prognostic and predictive value of tumor-infiltrating lymphocytes in a phase III randomized adjuvant breast cancer trial in node-positive breast cancer comparing the addition of docetaxel to doxorubicin with doxorubicin-based chemotherapy: BIG 02-98. J Clin Oncol 2013;31:860–867.

12 Adams S, Gray RJ, Demaria S, Goldstein L, Perez EA, Shulman LN, Martino S, Wang M, Jones VE, Saphner TJ, et al: Prognostic value of tumor-infiltrating lymphocytes in triple-negative breast cancers from two phase III randomized adjuvant breast cancer trials: ECOG 2197 and ECOG 1199. J Clin Oncol 2014; 32:2959–2966.

13 Demaria S, Volm MD, Shapiro RL, Yee HT, Oratz R, Formenti SC, Muggia F, Symmans WF: Development of tumor-infiltrating lymphocytes in breast cancer after neoadjuvant paclitaxel chemotherapy. Clin Cancer Res 2001;7:3025–3030.

14 Ladoire S, Mignot G, Dabakuyo S, Arnould L, Apetoh L, Rebe C, Coudert B, Martin F, Bizollon MH, Vanoli A, et al: In situ immune response after neoadjuvant chemotherapy for breast cancer predicts survival. J Pathol 2011;224:389–400.

15 Dieci MV, Criscitiello C, Goubar A, Viale G, Conte P, Guarneri V, Ficarra G, Mathieu MC, Delaloge S, Curigliano G, et al: Prognostic value of tumor-infiltrating lymphocytes on residual disease after primary chemotherapy for triple-negative breast cancer: a retrospective multicenter study. Ann Oncol 2014;25: 611–618.

16 Apetoh L, Ghiringhelli F, Tesniere A, Obeid M, Ortiz C, Criollo A, Mignot G, Maiuri MC, Ullrich E, Saulnier P, et al: Toll-like receptor 4-dependent contribution of the immune system to anticancer chemotherapy and radiotherapy. Nat Med 2007;13: 1050–1059.

17 Ruffell B, Au A, Rugo HS, Esserman LJ, Hwang ES, Coussens LM: Leukocyte composition of human breast cancer. Proc Natl Acad Sci USA 2012;109: 2796–2801.

18 Dedeurwaerder S, Desmedt C, Calonne E, Singhal SK, Haibe-Kains B, Defrance M, Michiels S, Volkmar M, Deplus R, Luciani J, et al: DNA methylation profiling reveals a predominant immune component in breast cancers. EMBO Mol Med 2011;3:726–741.
19 Andre F, Job B, Dessen P, Tordai A, Michiels S, Liedtke C, Richon C, Yan K, Wang B, Vassal G, et al: Molecular characterization of breast cancer with high-resolution oligonucleotide comparative genomic hybridization array. Clin Cancer Res 2009;15: 441–451.
20 Sotiriou C, Pusztai L: Gene-expression signatures in breast cancer. New Engl J Med 2009;360:790–800.
21 Reis-Filho JS, Weigelt B, Fumagalli D, Sotiriou C: Molecular profiling: moving away from tumor philately. Sci Transl Med 2010;2:47ps43.
22 Finak G, Bertos N, Pepin F, Sadekova S, Souleimanova M, Zhao H, Chen H, Omeroglu G, Meterissian S, Omeroglu A, et al: Stromal gene expression predicts clinical outcome in breast cancer. Nat Med 2008;14: 518–527.
23 Ignatiadis M, Singhal SK, Desmedt C, Haibe-Kains B, Criscitiello C, Andre F, Loi S, Piccart M, Michiels S, Sotiriou C: Gene modules and response to neoadjuvant chemotherapy in breast cancer subtypes: a pooled analysis. J Clin Oncol 2012;30:1996–2004.
24 Desmedt C, Haibe-Kains B, Wirapati P, Buyse M, Larsimont D, Bontempi G, Delorenzi M, Piccart M, Sotiriou C: Biological processes associated with breast cancer clinical outcome depend on the molecular subtypes. Clin Cancer Res 2008;14:5158–5165.
25 Teschendorff AE, Miremadi A, Pinder SE, Ellis IO, Caldas C: An immune response gene expression module identifies a good prognosis subtype in estrogen receptor negative breast cancer. Genome Biol 2007;8:R157.
26 Tan PH, Lota AS: Interaction of current cancer treatments and the immune system: implications for breast cancer therapeutics. Expert Opin Pharmacother 2008;9:2639–2660.
27 Zitvogel L, Apetoh L, Ghiringhelli F, Kroemer G: Immunological aspects of cancer chemotherapy. Nat Rev Immunol 2008;8:59–73.
28 Zitvogel L, Apetoh L, Ghiringhelli F, Andre F, Tesniere A, Kroemer G: The anticancer immune response: indispensable for therapeutic success? J Clin Invest 2008;118:1991–2001.
29 Lake RA, Robinson BW: Immunotherapy and chemotherapy – a practical partnership. Nat Rev 2005;5: 397–405.
30 van der Most RG, Currie AJ, Robinson BW, Lake RA: Decoding dangerous death: how cytotoxic chemotherapy invokes inflammation, immunity or nothing at all. Cell Death Differ 2008;15:13–20.
31 Obeid M, Tesniere A, Ghiringhelli F, Fimia GM, Apetoh L, Perfettini JL, Castedo M, Mignot G, Panaretakis T, Casares N, et al: Calreticulin exposure dictates the immunogenicity of cancer cell death. Nat Med 2007;13:54–61.
32 Casares N, Pequignot MO, Tesniere A, Ghiringhelli F, Roux S, Chaput N, Schmitt E, Hamai A, Hervas-Stubbs S, Obeid M, et al: Caspase-dependent immunogenicity of doxorubicin-induced tumor cell death. J Exp Med 2005;202:1691–1701.
33 Apetoh L, Ghiringhelli F, Tesniere A, Criollo A, Ortiz C, Lidereau R, Mariette C, Chaput N, Mira JP, Delaloge S, et al: The interaction between HMGB1 and TLR4 dictates the outcome of anticancer chemotherapy and radiotherapy. Immunol Rev 2007;220: 47–59.
34 Mattarollo SR, Loi S, Duret H, Ma Y, Zitvogel L, Smyth MJ: Pivotal role of innate and adaptive immunity in anthracycline chemotherapy of established tumors. Cancer Res 2011;71:4809–4820.
35 Zitvogel L, Kepp O, Kroemer G: Decoding cell death signals in inflammation and immunity. Cell 2010; 140:798–804.
36 Zitvogel L, Kepp O, Kroemer G: Immune parameters affecting the efficacy of chemotherapeutic regimens. Nat Rev Clin Oncol 2011;8:151–160.
37 Green DR, Ferguson T, Zitvogel L, Kroemer G: Immunogenic and tolerogenic cell death. Nat Rev Immunol 2009;9:353–363.
38 Periti P, Mini E: Immunomodulation by cancer chemotherapeutic agents. Chemioterapia 1987;6:399–402.
39 Sheard CR, Reilly F, Tee DE, Vergani D, Lowe D, Baum M, Cameron AE: The effect of adjuvant cyclophosphamide or tamoxifen on the numbers of lymphocytes bearing T cell or NK cell markers. Br J Cancer 1986;54:705–709.
40 Rozkova D, Horvath R, Bartunkova J, Spisek R: Glucocorticoids severely impair differentiation and antigen presenting function of dendritic cells despite upregulation of Toll-like receptors. Clin Immunol 2006;120:260–271.
41 Chiossone L, Vitale C, Cottalasso F, Moretti S, Azzarone B, Moretta L, Mingari MC: Molecular analysis of the methylprednisolone-mediated inhibition of NK-cell function: evidence for different susceptibility of IL-2- versus IL-15-activated NK cells. Blood 2007;109:3767–3775.
42 Schuerwegh AJ, van Offel JF, Bridts CH, Stevens WJ, De Clerck LS: Influence of longterm therapy with methotrexate and low dose corticosteroids on type 1 and type 2 cytokine production in CD4+ and CD8+ T lymphocytes of patients with rheumatoid arthritis. J Rheumatol 2001;28:1793–1799.

43 Schiavoni G, Mattei F, Di Pucchio T, Santini SM, Bracci L, Belardelli F, Proietti E: Cyclophosphamide induces type I interferon and augments the number of CD44^{hi} T lymphocytes in mice: implications for strategies of chemoimmunotherapy of cancer. Blood 2000;95:2024–2030.
44 Ghiringhelli F, Menard C, Puig PE, Ladoire S, Roux S, Martin F, Solary E, Le Cesne A, Zitvogel L, Chauffert B: Metronomic cyclophosphamide regimen selectively depletes CD4+CD25+ regulatory T cells and restores T and NK effector functions in end stage cancer patients. Cancer Immunol Immunother 2007;56:641–648.
45 Chakraborty M, Abrams SI, Coleman CN, Camphausen K, Schlom J, Hodge JW: External beam radiation of tumors alters phenotype of tumor cells to render them susceptible to vaccine-mediated T-cell killing. Cancer Res 2004;64:4328–4337.
46 Carson WE 3rd, Shapiro CL, Crespin TR, Thornton LM, Andersen BL: Cellular immunity in breast cancer patients completing taxane treatment. Clin Cancer Res 2004;10:3401–3409.
47 Machiels JP, Reilly RT, Emens LA, Ercolini AM, Lei RY, Weintraub D, Okoye FI, Jaffee EM: Cyclophosphamide, doxorubicin, and paclitaxel enhance the antitumor immune response of granulocyte/macrophage-colony stimulating factor-secreting whole-cell vaccines in HER-2/neu tolerized mice. Cancer Res 2001;61:3689–3697.
48 Yu B, Kusmartsev S, Cheng F, Paolini M, Nefedova Y, Sotomayor E, Gabrilovich D: Effective combination of chemotherapy and dendritic cell administration for the treatment of advanced-stage experimental breast cancer. Clin Cancer Res 2003;9:285–294.
49 Galetto A, Buttiglieri S, Forno S, Moro F, Mussa A, Matera L: Drug- and cell-mediated antitumor cytotoxicities modulate cross-presentation of tumor antigens by myeloid dendritic cells. Anticancer Drugs 2003;14:833–843.
50 Nowak AK, Robinson BW, Lake RA: Gemcitabine exerts a selective effect on the humoral immune response: implications for combination chemo-immunotherapy. Cancer Res 2002;62:2353–2358.
51 Suzuki E, Kapoor V, Jassar AS, Kaiser LR, Albelda SM: Gemcitabine selectively eliminates splenic Gr-1+/CD11b+ myeloid suppressor cells in tumor-bearing animals and enhances antitumor immune activity. Clin Cancer Res 2005;11:6713–6721.
52 Correale P, Aquino A, Giuliani A, Pellegrini M, Micheli L, Cusi MG, Nencini C, Petrioli R, Prete SP, De Vecchis L, et al: Treatment of colon and breast carcinoma cells with 5-fluorouracil enhances expression of carcinoembryonic antigen and susceptibility to HLA-A(*)02.01 restricted, CEA-peptide-specific cytotoxic T cells in vitro. Int J Cancer 2003;104:437–445.
53 Ghiringhelli F, Apetoh L, Tesniere A, Aymeric L, Ma Y, Ortiz C, Vermaelen K, Panaretakis T, Mignot G, Ullrich E, et al: Activation of the NLRP3 inflammasome in dendritic cells induces IL-1β-dependent adaptive immunity against tumors. Nat Med 2009; 15:1170–1178.
54 Sluyter R, Dalitz JG, Wiley JS: P2X7 receptor polymorphism impairs extracellular adenosine 5′-triphosphate-induced interleukin-18 release from human monocytes. Genes Immun 2004;5:588–591.
55 Brufsky A: Trastuzumab-based therapy for patients with HER2-positive breast cancer: from early scientific development to foundation of care. Am J Clin Oncol 2010;33:186–195.
56 Ross JS, Slodkowska EA, Symmans WF, Pusztai L, Ravdin PM, Hortobagyi GN: The HER-2 receptor and breast cancer: ten years of targeted anti-HER-2 therapy and personalized medicine. Oncologist 2009;14:320–368.
57 Musolino A, Naldi N, Bortesi B, Pezzuolo D, Capelletti M, Missale G, Laccabue D, Zerbini A, Camisa R, Bisagni G, et al: Immunoglobulin G fragment C receptor polymorphisms and clinical efficacy of trastuzumab-based therapy in patients with HER-2/neu-positive metastatic breast cancer. J Clin Oncol 2008; 26:1789–1796.
58 Tamura K, Shimizu C, Hojo T, Akashi-Tanaka S, Kinoshita T, Yonemori K, Kouno T, Katsumata N, Ando M, Aogi K, et al: FcγR2A and 3A polymorphisms predict clinical outcome of trastuzumab in both neoadjuvant and metastatic settings in patients with HER2-positive breast cancer. Ann Oncol 2011; 22:1302–1307.
59 Stagg J, Loi S, Divisekera U, Ngiow SF, Duret H, Yagita H, Teng MW, Smyth MJ: Anti-ErbB-2 mAb therapy requires type I and II interferons and synergizes with anti-PD-1 or anti-CD137 mAb therapy. Proc Natl Acad Sci USA 2011;108:7142–7147.
60 Park S, Jiang Z, Mortenson ED, Deng L, Radkevich-Brown O, Yang X, Sattar H, Wang Y, Brown NK, Greene M, et al: The therapeutic effect of anti-HER2/neu antibody depends on both innate and adaptive immunity. Cancer Cell 2010;18:160–170.
61 Rakhra K, Bachireddy P, Zabuawala T, Zeiser R, Xu L, Kopelman A, Fan AC, Yang Q, Braunstein L, Crosby E, et al: $CD4^+$ T cells contribute to the remodeling of the microenvironment required for sustained tumor regression upon oncogene inactivation. Cancer Cell 2010;18:485–498.
62 Brahmer JR, Drake CG, Wollner I, Powderly JD, Picus J, Sharfman WH, Stankevich E, Pons A, Salay TM, McMiller TL, et al: Phase I study of single-agent anti-programmed death-1 (MDX-1106) in refractory solid tumors: safety, clinical activity, pharmacodynamics, and immunologic correlates. J Clin Oncol 2010;28:3167–3175.

63 Hodi FS, O'Day SJ, McDermott DF, Weber RW, Sosman JA, Haanen JB, Gonzalez R, Robert C, Schadendorf D, Hassel JC, et al: Improved survival with ipilimumab in patients with metastatic melanoma. New Engl J Med 2010;363:711–723.
64 Persson J, Beyer I, Yumul R, Li Z, Kiem HP, Roffler S, Lieber A: Immuno-therapy with anti-CTLA4 antibodies in tolerized and non-tolerized mouse tumor models. PloS One 2011;6:e22303.
65 Curigliano G, Spitaleri G, Pietri E, Rescigno M, de Braud F, Cardillo A, Munzone E, Rocca A, Bonizzi G, Brichard V, Orlando L, Goldhirsch A: Breast cancer vaccines: a clinical reality or fairy tale? Ann Oncol 2006;17:750–762.
66 Rivoltini L, Canese P, Huber V, Iero M, Pilla L, Valenti R, Fais S, Lozupone F, Casati C, Castelli C, et al: Escape strategies and reasons for failure in the interaction between tumour cells and the immune system: how can we tilt the balance towards immune-mediated cancer control? Exp Opin Biol Ther 2005; 5:463–476.
67 Disis ML, Schiffman K, Guthrie K, Salazar LG, Knutson KL, Goodell V, dela Rosa C, Cheever MA: Effect of dose on immune response in patients vaccinated with an her-2/neu intracellular domain protein-based vaccine. J Clin Oncol 2004;22:1916–1925.
68 Goydos JS, Elder E, Whiteside TL, Finn OJ, Lotze MT: A phase I trial of a synthetic mucin peptide vaccine: induction of specific immune reactivity in patients with adenocarcinoma. J Surg Res 1996;63:298–304.
69 Knutson KL, Disis ML: Expansion of HER2/neu-specific T cells ex vivo following immunization with a HER2/neu peptide-based vaccine. Clin Breast Cancer 2001;2:73–79.
70 MacLean GD, Miles DW, Rubens RD, Reddish MA, Longenecker BM: Enhancing the effect of THERATOPE STn-KLH cancer vaccine in patients with metastatic breast cancer by pretreatment with low-dose intravenous cyclophosphamide. J Immunother Emphasis Tumor immunol 1996;19:309–316.
71 Marshall JL, Gulley JL, Arlen PM, Beetham PK, Tsang KY, Slack R, Hodge JW, Doren S, Grosenbach DW, Hwang J, et al: Phase I study of sequential vaccinations with fowlpox-CEA(6D)-TRICOM alone and sequentially with vaccinia-CEA(6D)-TRICOM, with and without granulocyte-macrophage colony-stimulating factor, in patients with carcinoembryonic antigen-expressing carcinomas. J Clin Oncol 2005; 23:720–731.
72 Emens LA: Chemotherapy and tumor immunity: an unexpected collaboration. Front Biosci 2008;13: 249–257.
73 Emens LA, Jaffee EM: Leveraging the activity of tumor vaccines with cytotoxic chemotherapy. Cancer Res 2005;65:8059–8064.
74 Finn OJ: Cancer immunology. New Engl J Med 2008; 358:2704–2715.
75 Pantel K, Brakenhoff RH, Brandt B: Detection, clinical relevance and specific biological properties of disseminating tumour cells. Nat Rev 2008;8:329–340.
76 Kang Y, Pantel K: Tumor cell dissemination: emerging biological insights from animal models and cancer patients. Cancer Cell 2013;23:573–581.
77 Hoos A, Parmiani G, Hege K, Sznol M, Loibner H, Eggermont A, Urba W, Blumenstein B, Sacks N, Keilholz U, et al: A clinical development paradigm for cancer vaccines and related biologics. J Immunother 2007;30:1–15.
78 Alexe G, Dalgin GS, Scanfeld D, Tamayo P, Mesirov JP, DeLisi C, Harris L, Barnard N, Martel M, Levine AJ, et al: High expression of lymphocyte-associated genes in node-negative HER2+ breast cancers correlates with lower recurrence rates. Cancer Res 2007; 67:10669–10676.
79 Rody A, Holtrich U, Pusztai L, Liedtke C, Gaetje R, Ruckhaeberle E, Solbach C, Hanker L, Ahr A, Metzler D, et al: T-cell metagene predicts a favorable prognosis in estrogen receptor-negative and HER2-positive breast cancers. Breast Cancer Res 2009; 11:R15.
80 Yau C, Esserman L, Moore DH, Waldman F, Sninsky J, Benz CC: A multigene predictor of metastatic outcome in early stage hormone receptor-negative and triple-negative breast cancer. Breast Cancer Res 2010;12:R85.

Giuseppe Curigliano, MD, PhD
Early Drug Development for Innovative Therapies Division
via Ripamonti, 435
IT–20141 Milan (Italy)
E-Mail giuseppe.curigliano@ieo.it

Michielin O, Coukos G (eds): Immuno-Oncology.
Prog Tumor Res. Basel, Karger, 2015, vol 42, pp 44–54 (DOI: 10.1159/000437184)

Current Developments in Actively Personalized Cancer Vaccination with a Focus on RNA as the Drug Format

Mustafa Diken[a, b] · Sebastian Kreiter[a, b] · Björn Kloke[b] · Ugur Sahin[a, b]

[a]TRON-Translational Oncology at the University Medical Center of the Johannes Gutenberg University, and [b]BioNTech RNA Pharmaceuticals GmbH, Mainz, Germany

Abstract

Developments in sequencing technologies have not only led to a rapid generation of genomic and transcriptional data from cancer patients, but also revealed the vast diversity of cancer-specific changes in patient tumors. Among these, mutation changes in the protein sequence can result in novel epitopes recognized by the immune system and, therefore, can be employed in the development of personalized vaccines. Thanks to its easy design and scalable GMP production, vaccines based on mRNAs coding for mutated epitopes have emerged as a reliable strategy for the exploitation of the potential of patient-specific genomic data. In this review, we provide an overview of recent developments in actively personalized vaccinations, with a special focus on the promise of mRNA vaccines.

Although major progress has been achieved in the 'war on cancer' that was declared more than 40 years ago by the US government, cancer still accounts for 1 in 4 of all deaths in industrialized countries according to the WHO's global burden of disease project. As our understanding of cancer biology has grown, we have learned that the interindividual heterogeneity between cancer patients is enormous. Since cancer is a disease caused by mutations, it is of high relevance for drug developers to comprehend the fact that less than 5% of mutations are shared between the tumors of different patients [1]. In analogy to the concept of the 'long tail', drug development in oncology also has to face the challenge of the individual profile that every single patient's tumor carries. Therefore, the medical need for the development of personalized cancer immunotherapy is apparent, and it seems logical to focus on mutations and the effects caused by them.

Mutations – A Valuable Source of Antigens

Cancer mutations can induce altered protein sequences [e.g. nonsynonymous single nucleotide variations (SNV), indels, fusions, splice site mutations], thereby potentially generating novel immunogenic epitopes. Just as the number of mutations varies in general, so too does the number of nonsynonymous mutations, by between tens and hundreds in differing cancer entities [1–12]. From the immunological perspective, protein-changing mutations may be particularly potent vaccine targets as they can create neoepitopes that are not subject to central tolerance mechanisms. Mutated peptides are absent in the thymus during development and are not endogenously expressed in normal tissue. Importantly, it has already been shown by Lennerz et al. [13] that natural antitumoral immunity in a melanoma patient is dominated by T cell responses against neoantigens. Cytotoxic T cells recognizing single amino acid changes originating from SNV have also been identified in other cancer entities. Wick et al. [14] found spontaneous CD8+ reactivity against a mutation in hydroxysteroid dehydrogenase-like protein-1 in 1 of 3 ovarian cancer patients. Interestingly, Schumacher et al. [15] also showed that mutation-specific CD4+ T cells occur spontaneously in glioma patients. Furthermore, there is accumulating data demonstrating a correlation between tumor rejection and the existence of T cell responses against mutated neoantigens. In 2013, Robbins and colleagues [16, 17] described that melanoma patients with objective responses after tumor-infiltrating lymphocyte (TIL) therapy included mutation-specific T cells in the transferred TILs. These data were reproduced and extended from the same group by studying 2 melanoma patients with durable complete responses, again demonstrating the targeting of mutated neoantigens by the transferred TILs [18]. Complementary to these results, in a melanoma patient responsive to ipilimumab, the induction of a neoantigen-specific CD8 TIL response could be traced [19]. The general focus on CD8+ T cells recognizing epitopes generated by SNV was challenged by a publication demonstrating that autologous CD4+ mutation-specific T cells adoptively transferred to a patient with metastatic cholangiocarcinoma could induce an objective clinical response [20]. An interesting and broader view is offered by an analysis of The Cancer Genome Association (TCGA) data for different tumor entities (colorectal, ovary, breast, brain, kidney, lung), which shows a significantly improved overall survival for those patients with at least one predicted immunogenic nonsynonymous mutation [21]. Previous studies have explored the potential immunogenicity of tumor mutations [22–24]. Segal et al. [22] predicted 40–60 histocompatibility leukocyte antigen (HLA) class I restricted epitopes per patient derived from SNV. Recently, Brown et al. [21] analyzed a set of 515 solid tumors for which they predicted, with a very stringent filter strategy, no immunogenic mutation in 334 patients and a median of 3 immunogenic mutations for the remaining 181. Concerning the vaccine design, it is important to consider the question whether immunoediting mechanisms might also act on mutated antigens. In a pivotal study, Schreiber and coworkers [25] showed that a mutated epitope can be lost in tumors by immunoediting. Therefore, in order to minimize the risk of such escaped mutants, it seems desirable to target more than one mutated neoepitope simultaneously.

Innovative Technology Enables Medical Progress in Personalized Therapies

Indeed, with the emergence of massively parallel sequencing technology, every patient's individual tumor mutanome 'signature' can be mapped within a short time frame in a cost-effective manner [26]. This technological breakthrough in individualized tumor genomic diagnostics now calls for exploitation with individualized therapeutic approaches. The advancement of sequencing technologies has revealed that human cancers carry dozens to hundreds of nonsynonymous mutations. Genome-wide mutation identification initiatives have been launched, including TCGA and the International Cancer Genome Consortium (ICGC). The work of Castle et al. [27] provided important support for the use of DNA-based mutation-detection assays for patient stratification by showing that mutations are transcribed into messenger (m)RNA in equal proportions to their allele frequency. Novel concepts are required to translate the comprehensive molecular cancer map of a patient to an on-demand, GMP (good manufacturing practice), patient-specific therapeutic. Further progress with respect to predicting the immunogenicity of mutations is highly relevant to this process. Substantial advancements in recent years has led to the establishment of versatile algorithms which can be used to predict the position of possible antigenic epitopes [28–30]. Further advances, especially in the prediction of HLA class II restricted peptides, will foster the clinical success of actively personalized cancer vaccinations.

The Promise of Cancer Vaccines: Focus on mRNA as an Emerging Format

Exploitation of the genomic data for the development of patient-specific drugs is of great importance and RNA-based cancer vaccines can serve as an appropriate drug format for this purpose. Indeed, it has been more than 1 century since the initial attempt was made to harness the potential of our immune system in the fight against cancer in a nonpersonalized manner. In 1893, William B. Coley [31] was the first to show that treatment of cancer patients through unspecific stimulation of the immune system with a mixture of attenuated bacteria was principally feasible. His first nontumor-originated class of vaccine consisting of an ill-defined crude mixture has now evolved into a molecularly defined new generation of vaccines carrying well-characterized antigens of tumor origin.

These tumor antigens are currently being introduced into various vaccine formats which all aim to generate antigen-specific immune responses through the priming of T cells via antigen-presenting cells that engulf the vaccine-carried antigen. One of the early approaches for antigen-specific immunotherapy was to use short or long peptides derived from tumor antigens and a suitable adjuvant [32]. Although the synthesis of these peptides is feasible, immune responses against peptide vaccines are restricted to selected patients that possess the appropriate HLA haplotype, thus limiting their use in broader patient populations. The use of complete protein antigen solved the problem of HLA restriction as the protein contains the complete repertoire of antigenic peptides. However, this strategy suffers from the complex and expensive vaccine-grade production of recombinant proteins and their limited immunogenicity.

Vector vaccines are yet another strategy to deliver the antigenic information as naked or virus-carried nucleic acid. Early studies employed DNA or viral vaccines with the risk of genomic integration and unwanted immunogenicity against the DNA itself or virus-derived antigens, thus presenting safety concerns. On the other hand, use of mRNA as a drug format for vaccination purposes has become a promising approach in recent years, compensating for the drawbacks of other vector and nonvector vaccine formats [33]. The mRNA is only transiently expressed in cells lasting for a few days, which is enough for the induction of immunity and mRNA does not get integrated into the host's genome. Moreover, in contrast to peptide vaccination in which the HLA haplotype of the patient must be prior determined, vaccination with mRNA encoding the whole antigen obviates this requirement. mRNA, through being recognized by pattern-recognition receptors, can additionally serve as a vaccine adjuvant, obviating the need for an additional adjuvant as in the case of peptide vaccinations. Such mRNA vaccines with antigen-encoding and immunostimulatory properties can be easily produced via in vitro transcription in a cost-effective manner and tailored in order to exploit its potential for research and clinical use [34].

In 1995, naked mRNA was first used for the induction of antigen-specific immune responses in mice upon intramuscular injection of mRNA encoding for carcinoembryonic antigen [35]. Similarly, intradermal injection of naked β-galactosidase mRNA to the ear pinna of mice resulted in the induction of cytotoxic T cells as well as antibody responses against the encoded protein [36]. The combination of GM-CSF treatment with naked mRNA vaccination in a similar setting enhanced the humoral response and further polarized it to cellular immune responses [37]. Intradermal delivery of mRNA upon its complexation with protamine is also followed in various studies to protect RNA from RNase-mediated degradation. Initial studies involved the direct administration of protamine-complexed RNA and were able to induce tumor regression and long-term antitumor immunity [38]. A newer generation of such vaccines was comprised of two components, free and protamine-complexed mRNA, to balance the protein expression and immunostimulation, respectively [39]. These vaccines protected mice from tumor challenge and elicited therapeutic effects in preclinical mouse models of cancer. Moreover, the combination of such mRNA vaccines with chemotherapy, radiotherapy and emerging immune check-point antibodies, such as anti-CTLA-4 antibody, was feasible and improved their therapeutic potential [40, 41].

Promising results from preclinical studies encouraged clinical testing of the intradermal delivery of mRNA vaccines. The first study employed a naked tumor mRNA library injected directly to stage III and IV grade melanoma patients using GM-CSF as the adjuvant. Although no objective clinical responses were observed, more than half of the patients exhibited antivaccine T cell and antibody responses [42]. In yet another clinical trial in metastatic melanoma, mRNA-protamine complexes encoding for tumor-associated antigens like Melan A, Mage A1, Mage A3, tyrosinase, gp100 and survivin were administered intradermally together with GM-CSF and led to antigen-specific immune responses as well as a decrease in the metastatic load [43]. The potency of intradermal mRNA delivery was also tested in renal cell carcinoma with mRNAs encoding for mucin 1, carcinoembryonic antigen, telomerase, survivin, melanoma-associated antigen 1 and human epidermal growth factor receptor-2 (Her-2/neu). In half of the patients, clinical benefit defined as tumor regression or stabilization lasting for >3 months during administration of the vaccine was observed [44]. Two-component mRNA vaccines are also being tested in various clinical trials with indications of prostate and non-small cell lung cancers, the results of which are currently awaited (NCT01915524, NCT01817738, NCT00831467).

Direct intranodal injection of naked antigen-encoding mRNA is yet another strategy which has reached clinical testing. Lymph node-resident antigen-presenting cells can be directly targeted by this approach obviating the need for the migration of these cells from the periphery to lymph nodes upon mRNA internalization. It has been shown that mRNA administered intranodally is selectively engulfed by lymph node-resident dendritic cells via micropinocytosis [45], leading to their activation and propagation of a T cell-attracting milieu in the lymph node [46]. Intranodal vaccination with pharmacologically optimized naked mRNA with enhanced translational and T cell stimulatory properties [47–49] was able to elicit potent prophylactic and therapeutic antitumoral immunity in preclinical models and, moreover, was superior to other immunization routes [46]. The potency of the intranodal vaccine could be improved by its rational combination with the Fms-like tyrosine kinase 3 ligand, a naturally occurring growth factor for dendritic cells that causes their expansion in the lymph node [50]. Another effective combination of intranodal mRNA vaccination in the preclinical setting was achieved via rapamycin. Use of this mTOR inhibitor, which is primarily used for immunosuppression in the clinic, upon mRNA vaccination at the effector-to-memory transition phase led to the generation of a superior T cell memory pool mirrored by improved survival in mice [51]. In addition to non-RNA adjuvants, mRNAs encoding for immunostimulatory molecules like CD40, CD70 and constitutively active TLR4 were also able to enhance the therapeutic potential of mRNA vaccines in several mouse tumor models when combined with the antigen-encoding mRNA [52]. The intranodal delivery of naked antigen-encoding RNA can be performed easily under ultrasound guidance in the clinical setting and its potency is currently being tested in melanoma patients (NCT01684241). Independent of the route of administration, direct mRNA vaccination was shown to be safe in both preclinical and clinical settings without serious adverse effects and holds promise as a vaccination strategy against cancer.

Preclinical Proof of Concept for Personalized Vaccination

Pursuing for the first time an experimental proof for the concept of actively personalized cancer vaccination, we combined high-coverage deep sequencing of the B16F10 melanoma with systematic immunogenicity analyses of the identified mutations. B16F10 is a very aggressive melanoma cell line clone which spontaneously originated in a C57BL/6 mouse [53] and is regarded as poorly immunogenic due to the low major histocompatibility complex (MHC) expression [54]. We identified 3,570 somatic mutations, including 1,392 in transcripts, of which 1,266 were in protein coding regions. Of these, 962 cause nonsynonymous protein changes and, of these, 562 occur in expressed genes based on RNA-Seq data [55]. Beyond B16, we turned to the murine colorectal cancer cell line CT26, which is one of the most commonly used cell lines in drug development. We utilized NGS (next-generation sequencing), bioinformatics and immunoinformatics to create an integrated mouse solid tumor mutanome, transcriptome and immunome. The mutations (1,688 SNV causing nonsynonymous protein changes) and expression profiles are similar to those reported for sporadic, undifferentiated, therapy refractory, metastasis-prone human colorectal cancer [56]. The immunogenicity and specificity of 50 selected and validated B16 mutations were assessed by immunizing mice with long synthesized peptides encoding the mutated epitope. The long peptides were 27-amino acids in length with the mutated or wild-type amino acid positioned centrally. Approximately one third (16) of the 50 mutant peptides were found to be immunogenic, with 60% (11) of this immunogenic group eliciting immune responses directed preferentially against the mutated sequence as compared to the wild-type sequence. For peptides inducing a strong mutation-specific T cell response, immune recognition was confirmed by an independent approach proving that induced, mutation-specific T cells recognize endogenously processed epitopes. Significantly, the study demonstrated the immunogenicity of novel somatic mutations that are not known to promote a tumor phenotype, thus indicating that, regardless of their function, immunity against these mutations, as long as they are stably expressed in a relevant malignant cell population, may mediate tumor control. Finally, we showed that vaccination with mutated epitopes results in significant tumor growth retardation. Further studies proved that mRNA-encoded neoantigens can be used to confer antitumoral immunity in different murine tumor models [(CT26, 4T-1), in preparation]. This was possible not only with RNA-encoding single mutated epitopes, but also utilizing polytopic RNA constructs tested in different murine cancer models.

Clinical Testing of an Actively Personalized mRNA Vaccination

In December 2013, BioNTech RNA Pharmaceuticals was able to start a 'first in men' phase I study, moving the concept of actively personalized mRNA vaccination into clinical testing (NCT02035956). The IVAC MUTANOME vaccine approach is based on targeting multiple immunogenic tumor mutations unique to a given patient's tumor and addressing cancer heterogeneity, and is therefore in principle applicable to all cancers carrying multiple mutations. The patients receive two separate polypeptide-encoding mRNA molecules manufactured for use in a single patient only, also termed the IVAC MUTANOME vaccine. Each of the two formulated RNAs encodes five mutation-coding sequences (Pentaepitope) that are unique for a given patient. The encoded mutation-coding sequences are identified through sequence profiling of a patient's healthy and tumor tissues, prioritized by bioinformatic algorithms, and finally confirmed by Sanger sequencing. Prioritization relies on the confidence of the identified somatic mutation, expression of the relevant gene, change of the amino acid sequence of the respective protein and predicted MHC presentation, as well as immune cell recognition of the potential neoepitope. Both drug products are manufactured under GMP for each patient individually and applied by ultrasound-guided intralymphatic injection. Parallel to the target discovery process and on-demand manufacturing of the IVAC MUTANOME vaccine, patients with NY-ESO-1 and/or tyrosinase-positive tumors will receive an off-the-shelf mRNA vaccination targeting both tumor-associated antigens. These vaccines were introduced to clinical testing in a phase I clinical trial (NCT01684241) in 2012 and have been shown to be well tolerated in patients following repetitive intranodal injections with up to 1,000 μg of mRNA.

Currently 8 patients have been enrolled into the IVAC MUTANOME clinical trial. The aim of the first clinical trial to test the IVAC MUTANOME concept is to determine the safety, tolerability and immunogenicity of the personalized vaccination approach for malignant melanoma. The first 3 patients have been treated and for the other 5 patients manufacturing of the IVAC MUTANOME vaccine is ongoing. So far no severe adverse effects or treatment-related adverse events have been reported and, therefore, the IVAC MUTANOME vaccine shows good safety and tolerability. The IVAC MUTANOME trial is the first in Europe that introduces a fully personalized mutanome vaccine in malignant melanoma. The first proof-of-concept results are expected in 2015.

Outlook and Challenges

Personalized cancer vaccination therapy has just recently entered clinical translation and many challenges have to be overcome before it can become a standard of care treatment modality. Scientifically, it will be important to improve the bioinformatic prediction of immunogenicity for missense mutations in order to increase the vaccine potency. The inclusion of indels and fusions will eventually open up further sources of neoantigens to be utilized in vaccine design. Conceptually, additional complexity is added by the fact that intratumoral heterogeneity poses a further challenge to any upcoming immunotherapeutic approach. Additional efforts will therefore be necessary to develop the optimal vaccine design. Thanks to the advantages mentioned in this review, mRNA-based vaccines hold promise for exploiting the potential of personalized therapies. To guarantee patient safety and also foster clinical translation a dialogue between developers and regulators mediated by the Association for Cancer Immunotherapy (CIMT) concerning the fitting regulatory framework has already been started and is currently ongoing [57]. Finally, the high costs of personalized cancer vaccination pose a significant developmental risk. Decreasing the cost of NGS and full automation, as well as optimization of the manufacturing process and scale effects, will probably facilitate the approval of affordable products.

References

1 Stratton MR: Exploring the genomes of cancer cells: progress and promise. Science 2011;331:1553–1558.
2 Greenman C, Stephens P, Smith R, et al: Patterns of somatic mutation in human cancer genomes. Nature 2007;446:153–158.
3 Ding L, Ellis MJ, Li S, et al: Genome remodelling in a basal-like breast cancer metastasis and xenograft. Nature 2010;464:999–1005.
4 Parsons DW, Jones S, Zhang X, et al: An integrated genomic analysis of human glioblastoma multiforme. Science 2008;321:1807–1812.
5 Jones S, Zhang X, Parsons DW, et al: Core signaling pathways in human pancreatic cancers revealed by global genomic analyses. Science 2008;321:1801–1806.
6 Totoki Y, Tatsuno K, Yamamoto S, Arai Y, Hosoda F, Ishikawa S, Tsutsumi S, Sonoda K, Totsuka H, Shirakihara T, Sakamoto H, Wang L, Ojima H, Shimada K, Kosuge T, Okusaka T, Kato K, Kusuda J, Yoshida T, Aburatani H, Shibata T: High-resolution characterization of a hepatocellular carcinoma genome. Nat Genet 2011;43:464–469.
7 Sjoblom T, Jones S, Wood LD, Parsons DW, Lin J, Barber TD, Mandelker D, Leary RJ, Ptak J, Silliman N, Szabo S, Buckhaults P, Farrell C, Meeh P, Markowitz SD, Willis J, Dawson D, Willson JK, Gazdar AF, Hartigan J, Wu L, Liu C, Parmigiani G, Park BH, Bachman KE, Papadopoulos N, Vogelstein B, Kinzler KW, Velculescu VE: The consensus coding sequences of human breast and colorectal cancers. Science 2006;314:268–274.
8 Wood LD, Parsons DW, Jones S, et al: The genomic landscapes of human breast and colorectal cancers. Science 2007;318:1108–1113.
9 Pleasance ED, Stephens PJ, O'Meara S, et al: A small-cell lung cancer genome with complex signatures of tobacco exposure. Nature 2010;463:184–190.
10 Pleasance ED, Cheetham RK, Stephens PJ, McBride DJ, et al: A comprehensive catalogue of somatic mutations from a human cancer genome. Nature 2010;463:191–196.
11 Wei X, Walia V, Lin JC, Teer JK, Prickett TD, Gartner J, Davis S, Stemke-Hale K, Davies MA, Gershenwald JE, Robinson W, Robinson S, Rosenberg SA, Samuels Y: Exome sequencing identifies GRIN2A as frequently mutated in melanoma. Nat Genet 2011;43:442–446.

12 Lee W, Jiang Z, Liu J, Haverty PM, Guan Y, Stinson J, Yue P, Zhang Y, Pant KP, Bhatt D, Ha C, Johnson S, Kennemer MI, Mohan S, Nazarenko I, Watanabe C, Sparks AB, Shames DS, Gentleman R, de Sauvage FJ, Stern H, Pandita A, Ballinger DG, Drmanac R, Modrusan Z, Seshagiri S, Zhang Z: The mutation spectrum revealed by paired genome sequences from a lung cancer patient. Nature 2010;465:473–477.
13 Lennerz V, Fatho M, Gentilini C, Frye RA, Lifke A, Ferel D, Wolfel C, Huber C, Wolfel T: The response of autologous T cells to a human melanoma is dominated by mutated neoantigens. Proc Natl Acad Sci U S A 2005;102:16013–16018.
14 Wick DA, Webb JR, Nielsen JS, Martin SD, Kroeger DR, Milne K, Castellarin M, Twumasi-Boateng K, Watson PH, Holt RA, Nelson BH: Surveillance of the tumor mutanome by T cells during progression from primary to recurrent ovarian cancer. Clin Cancer Res 2014;20:1125–1134.
15 Schumacher T, Bunse L, Pusch S, Sahm F, Wiestler B, Quandt J, Menn O, Osswald M, Oezen I, Ott M, Keil M, Balss J, Rauschenbach K, Grabowska AK, Vogler I, Diekmann J, Trautwein N, Eichmuller SB, Okun J, Stevanovic S, Riemer AB, Sahin U, Friese MA, Beckhove P, von Deimling A, Wick W, Platten M: A vaccine targeting mutant IDH1 induces antitumour immunity. Nature 2014;512:324–327.
16 Robbins PF, Lu YC, el-Gamil M, Li YF, Gross C, Gartner J, Lin JC, Teer JK, Cliften P, Tycksen E, Samuels Y, Rosenberg SA: Mining exomic sequencing data to identify mutated antigens recognized by adoptively transferred tumor-reactive T cells. Nat Med 2013;19:747–752.
17 Lu YC, Yao X, Li YF, el-Gamil M, Dudley ME, Yang JC, Almeida JR, Douek DC, Samuels Y, Rosenberg SA, Robbins PF: Mutated PPP1R3B is recognized by T cells used to treat a melanoma patient who experienced a durable complete tumor regression. J Immunol 2013;190:6034–6042.
18 Lu YC, Yao X, Crystal JS, Li YF, el-Gamil M, Gross C, Davis L, Dudley ME, Yang JC, Samuels Y, Rosenberg SA, Robbins PF: Efficient identification of mutated cancer antigens recognized by T cells associated with durable tumor regressions. Clin Cancer Res 2014;20:3401–3410.
19 van RN, van Buuren MM, Philips D, Velds A, Toebes M, Heemskerk B, van Dijk LJ, Behjati S, Hilkmann H, El AD, Nieuwland M, Stratton MR, Kerkhoven RM, Kesmir C, Haanen JB, Kvistborg P, Schumacher TN: Tumor exome analysis reveals neoantigen-specific T-cell reactivity in an ipilimumab-responsive melanoma. J Clin Oncol 2013;31:e439–e442.
20 Tran E, Turcotte S, Gros A, Robbins PF, Lu YC, Dudley ME, Wunderlich JR, Somerville RP, Hogan K, Hinrichs CS, Parkhurst MR, Yang JC, Rosenberg SA: Cancer immunotherapy based on mutation-specific CD4+ T cells in a patient with epithelial cancer. Science 2014;344:641–645.
21 Brown SD, Warren RL, Gibb EA, Martin SD, Spinelli JJ, Nelson BH, Holt RA: Neo-antigens predicted by tumor genome meta-analysis correlate with increased patient survival. Genome Res 2014;24:743–750.
22 Segal NH, Parsons DW, Peggs KS, Velculescu V, Kinzler KW, Vogelstein B, Allison JP: Epitope landscape in breast and colorectal cancer. Cancer Res 2008;68:889–892.
23 Warren RL, Holt RA: A census of predicted mutational epitopes suitable for immunologic cancer control. Hum Immunol 2010;71:245–254.
24 Khalili JS, Hanson RW, Szallasi Z: In silico prediction of tumor antigens derived from functional missense mutations of the cancer gene census. Oncoimmunology 2012;1:1281–1289.
25 Matsushita H, Vesely MD, Koboldt DC, Rickert CG, Uppaluri R, Magrini VJ, Arthur CD, White JM, Chen YS, Shea LK, Hundal J, Wendl MC, Demeter R, Wylie T, Allison JP, Smyth MJ, Old LJ, Mardis ER, Schreiber RD: Cancer exome analysis reveals a T-cell-dependent mechanism of cancer immunoediting. Nature 2012;482:400–404.
26 Koboldt DC, Steinberg KM, Larson DE, Wilson RK, Mardis ER: The next-generation sequencing revolution and its impact on genomics. Cell 2013;155:27–38.
27 Castle JC, Loewer M, Boegel S, Tadmor AD, Boisguerin V, de Graaf J, Paret C, Diken M, Kreiter S, Tureci O, Sahin U: Mutated tumor alleles are expressed according to their DNA frequency. Sci Rep 2014;4:4743.
28 Zhang Q, Wang P, Kim Y, Haste-Andersen P, Beaver J, Bourne PE, Bui HH, Buus S, Frankild S, Greenbaum J, Lund O, Lundegaard C, Nielsen M, Ponomarenko J, Sette A, Zhu Z, Peters B: Immune epitope database analysis resource (IEDB-AR). Nucleic Acids Res 2008;36:W513–W518.
29 Lundegaard C, Lamberth K, Harndahl M, Buus S, Lund O, Nielsen M: NetMHC-3.0: accurate web accessible predictions of human, mouse and monkey MHC class I affinities for peptides of length 8–11. Nucleic Acids Res 2008;36:W509–W512.
30 Karosiene E, Rasmussen M, Blicher T, Lund O, Buus S, Nielsen M: NetMHCIIpan-3.0, a common pan-specific MHC class II prediction method including all three human MHC class II isotypes, HLA-DR, HLA-DP and HLA-DQ. Immunogenetics 2013;65: 711–724.

31 Coley WB: The treatment of malignant tumors by repeated inoculations of erysipelas. With a report of ten original cases. 1893. Clin Orthop Relat Res 1991; 262:3–11.

32 Mandelboim O, Vadai E, Fridkin M, Katz-Hillel A, Feldman M, Berke G, Eisenbach L: Regression of established murine carcinoma metastases following vaccination with tumour-associated antigen peptides. Nat Med 1995;1:1179–1183.

33 Sahin U, Kariko K, Tureci O: mRNA-based therapeutics – developing a new class of drugs. Nat Rev Drug Discov 2014;30:759–780.

34 Pascolo S: Vaccination with messenger RNA (mRNA). Handb Exp Pharmacol 2008;183:221–235.

35 Conry RM, LoBuglio AF, Wright M, Sumerel L, Pike MJ, Johanning F, Benjamin R, Lu D, Curiel DT: Characterization of a messenger RNA polynucleotide vaccine vector. Cancer Res 1995;55:1397–1400.

36 Hoerr I, Obst R, Rammensee HG, Jung G: In vivo application of RNA leads to induction of specific cytotoxic T lymphocytes and antibodies. Eur J Immunol 2000;30:1–7.

37 Carralot JP, Probst J, Hoerr I, Scheel B, Teufel R, Jung G, Rammensee HG, Pascolo S: Polarization of immunity induced by direct injection of naked sequence-stabilized mRNA vaccines. Cell Mol Life Sci 2004;61:2418–2424.

38 Scheel B, Aulwurm S, Probst J, Stitz L, Hoerr I, Rammensee HG, Weller M, Pascolo S: Therapeutic antitumor immunity triggered by injections of immunostimulating single-stranded RNA. Eur J Immunol 2006;36:2807–2816.

39 Fotin-Mleczek M, Duchardt KM, Lorenz C, Pfeiffer R, Ojkic-Zrna S, Probst J, Kallen KJ: Messenger RNA-based vaccines with dual activity induce balanced TLR-7 dependent adaptive immune responses and provide antitumor activity. J Immunother 2011;34:1–15.

40 Fotin-Mleczek M, Zanzinger K, Heidenreich R, Lorenz C, Thess A, Duchardt KM, Kallen KJ: Highly potent mRNA based cancer vaccines represent an attractive platform for combination therapies supporting an improved therapeutic effect. J Gene Med 2012;14:428–439.

41 Fotin-Mleczek M, Zanzinger K, Heidenreich R, Lorenz C, Kowalczyk A, Kallen KJ, Huber SM: mRNA-based vaccines synergize with radiation therapy to eradicate established tumors. Radiat Oncol 2014;9:180.

42 Weide B, Carralot JP, Reese A, Scheel B, Eigentler TK, Hoerr I, Rammensee HG, Garbe C, Pascolo S: Results of the first phase I/II clinical vaccination trial with direct injection of mRNA. J Immunother 2008; 31:180–188.

43 Weide B, Pascolo S, Scheel B, Derhovanessian E, Pflugfelder A, Eigentler TK, Pawelec G, Hoerr I, Rammensee HG, Garbe C: Direct injection of protamine-protected mRNA: results of a phase 1/2 vaccination trial in metastatic melanoma patients. J Immunother 2009;32:498–507.

44 Rittig SM, Haentschel M, Weimer KJ, Heine A, Muller MR, Brugger W, Horger MS, Maksimovic O, Stenzl A, Hoerr I, Rammensee HG, Holderried TA, Kanz L, Pascolo S, Brossart P: Intradermal vaccinations with RNA coding for TAA generate CD8+ and CD4+ immune responses and induce clinical benefit in vaccinated patients. Mol Ther 2011;19:990–999.

45 Diken M, Kreiter S, Selmi A, Britten CM, Huber C, Tureci O, Sahin U: Selective uptake of naked vaccine RNA by dendritic cells is driven by macropinocytosis and abrogated upon DC maturation. Gene Ther 2011;18:702–708.

46 Kreiter S, Selmi A, Diken M, Koslowski M, Britten CM, Huber C, Tureci O, Sahin U: Intranodal vaccination with naked antigen-encoding RNA elicits potent prophylactic and therapeutic antitumoral immunity. Cancer Res 2010;70:9031–9040.

47 Holtkamp S, Kreiter S, Selmi A, Simon P, Koslowski M, Huber C, Tureci O, Sahin U: Modification of antigen-encoding RNA increases stability, translational efficacy, and T-cell stimulatory capacity of dendritic cells. Blood 2006;108:4009–4017.

48 Kreiter S, Selmi A, Diken M, Sebastian M, Osterloh P, Schild H, Huber C, Tureci O, Sahin U: Increased antigen presentation efficiency by coupling antigens to MHC class I trafficking signals. J Immunol 2008; 180:309–318.

49 Kuhn AN, Diken M, Kreiter S, Selmi A, Kowalska J, Jemielity J, Darzynkiewicz E, Huber C, Tureci O, Sahin U: Phosphorothioate cap analogs increase stability and translational efficiency of RNA vaccines in immature dendritic cells and induce superior immune responses in vivo. Gene Ther 2010;17:961–971.

50 Kreiter S, Diken M, Selmi A, Diekmann J, Attig S, Husemann Y, Koslowski M, Huber C, Tureci O, Sahin U: FLT3 ligand enhances the cancer therapeutic potency of naked RNA vaccines. Cancer Res 2011; 71:6132–6142.

51 Diken M, Kreiter S, Vascotto F, Selmi A, Attig S, Diekmann J, Huber C, Tureci O, Sahin U: mTOR inhibition improves antitumor effects of vaccination with antigen-encoding RNA. Cancer Immunol Res 2013;1:386–392.

52 van Lint S, Goyvaerts C, Maenhout S, Goethals L, Disy A, Benteyn D, Pen J, Bonehill A, Heirman C, Breckpot K, Thielemans K: Preclinical evaluation of TriMix and antigen mRNA-based antitumor therapy. Cancer Res 2012;72:1661–1671.

53 Fidler IJ: Selection of successive tumour lines for metastasis. Nat New Biol 1973;242:148–149.

54 Boegel S, Lower M, Schafer M, Bukur T, de Graaf J, Boisguerin V, Tureci O, Diken M, Castle JC, Sahin U: HLA typing from RNA-Seq sequence reads. Genome Med 2013;4:102.
55 Castle JC, Kreiter S, Diekmann J, Lower M, van de Roemer N, de Graaf J, Selmi A, Diken M, Boegel S, Paret C, Koslowski M, Kuhn AN, Britten CM, Huber C, Tureci O, Sahin U: Exploiting the mutanome for tumor vaccination. Cancer Res 2012;72:1081–1091.
56 Castle JC, Loewer M, Boegel S, de Graaf J, Bender C, Tadmor AD, Boisguerin V, Bukur T, Sorn P, Paret C, Diken M, Kreiter S, Tureci O, Sahin U: Immunomic, genomic and transcriptomic characterization of CT26 colorectal carcinoma. BMC Genomics 2014; 15:190.
57 Britten CM, Singh-Jasuja H, Flamion B, Hoos A, Huber C, Kallen KJ, Khleif SN, Kreiter S, Nielsen M, Rammensee HG, Sahin U, Hinz T, Kalinke U: The regulatory landscape for actively personalized cancer immunotherapies. Nat Biotechnol 2013;31:880–882.

Sebastian Kreiter, MD
Langenbeckstrasse 1
Building 708, TRON
DE–55131 Mainz (Germany)
E-Mail sebastian.kreiter@tron-mainz.de

Michielin O, Coukos G (eds): Immuno-Oncology.
Prog Tumor Res. Basel, Karger, 2015, vol 42, pp 55–66 (DOI: 10.1159/000437178)

Immune Checkpoint Inhibitors

John B.A.G. Haanen[a] · Caroline Robert[b]

[a]Divisions of Medical Oncology and Immunology, The Netherlands Cancer Institute, Amsterdam, The Netherlands; [b]Gustave Roussy, Service de Dermatologie et Université Paris-Sud, Villejuif-Paris, France

Abstract

Undoubtedly the discovery of immune checkpoints such as CTLA-4 and PD-1 has been crucial to the development of cancer immunotherapy. Although these molecules were originally discovered as molecules playing a role in T cell activation or apoptosis, subsequent preclinical research showed their important role in the maintenance of peripheral immune tolerance. Mice deficient of the immune checkpoints CTLA-4 or PD-1 develop autoimmune-like diseases that occur early after birth and are lethal in the case of CTLA-4 deficiency, or become apparent much later in life in the case of PD-1 deficiency. Blockade of CTLA-4 and PD-1 resulted in the development of antitumor immune responses that were effective as single agents or required additional treatment depending on the preclinical model. Therefore, it was surprising that single-agent anti-CTLA-4 and anti-PD-1 are so effective anticancer treatments in humans. These therapies have revolutionized cancer immunotherapy as they showed for the first time in many years of research in metastatic melanoma, which is considered one of the most immunogenic human cancers, an improvement in overall survival, with an increasing group of patients benefitting long-term from these treatments. In this chapter we discuss the discovery of immune checkpoints, the clinical application of their inhibitors and the future directions of this highly interesting class of molecules.

CTLA-4

CTLA-4 (cytotoxic T lymphocyte antigen-4) was first discovered in a cytotoxic T lymphocyte cDNA library as a molecule belonging to the immunoglobulin superfamily, which was expressed mainly by activated T cells [1]. Although CTLA-4 binds to the same ligand B7 on B cells and antigen-presenting cells as its homologue CD28, stimulation of CTLA-4 did not result in T cell activation, but led to T cell-mediated inhibition of antibody production and prevention of allograft rejection [2, 3]. Immunocompetent mice transplanted with HPV E7-expressing tumors rejected these tumors when cotransfected with the costimulatory molecule B7. The idea was that through C28/CTLA-4 costimulation of E7-specific CD8 T cells an endogenous antitumor immune response was induced that could even reject B7-negative tumors at distant sites [4]. At that time much less was known about the differential roles of CD28 and CTLA-4 in costimulation. In 1994 it was found that the kinetics of CTLA-4 expression was quite different from CD28 expression. Whereas CD28 is expressed on naïve T cells, CTLA-4 expression is induced for 2–3 days after T cell receptor (TCR)/CD3-mediated T cell stimulation, starting around 24 h following TCR triggering. Blockade of CTLA-4-B7 interaction by monoclonal anti-CTLA-4 antibodies led to potentiation of an alloantigen response, which was interpreted as the inhibition of delivering a negative signal to the T cell [5]. Therefore, whereas CD28 ligation is important for full T cell activation, CTLA-4 triggering results in an inhibitory signal upon TCR-mediated T cell activation. This role of CTLA-4 was corroborated in vivo by the generation of CTLA-4-deficient mice [6]. These mice develop normally until birth, but then display severe lymphoproliferative disease with up to a 10-fold enlargement of lymphoid organs (thymus, spleen and lymph nodes) and diffuse infiltration of lymphocytes in many organs, including the myocardium. Typically these animals succumb at the age of 3–4 weeks. T cells from these animals show a highly activated phenotype and proliferate more abundantly to TCR/CD3 triggering than T cells from normal littermates. CD4 and CD8 T cell numbers were greatly increased, but their ratio was unchanged. These results indicate that CTLA-4 plays an important role in the regulation of activated T cells, as lack of CTLA-4 leads to uncontrolled T cell proliferation. These insights in the mechanism of action of CTLA-4 led to the idea of investigating whether CTLA-4 blockade could improve antitumor immune responses. In a first study performed by Allison and colleagues [7], CTLA-4 blockade using a blocking monoclonal antibody prevented the outgrowth of both a transplantable murine colon carcinoma and murine sarcoma. Importantly, established tumors were rejected upon anti-CTLA-4 treatment and mice that had rejected tumors were protected from the outgrowth of tumors upon secondary challenge. However, in highly tumorigenic but poorly immunogenic tumors such as the murine melanoma B16, anti-CTLA-4 was hardly effective as a single agent. Combined with a GM-CSF-expressing tumor cell vaccine, the cure rate could be improved from 16% for the vaccine alone to 80% for the combination [8]. Interestingly, mice that effectively rejected the tumor developed skin depigmentation as a result of induction of a T cell response against the tyrosinase-related protein-2, TRP-2 [9]. Based on these compelling results, Hodi et al. [10] performed a phase I study with a single infusion of the CTLA-4 blocking antibody MDX-CTLA4 in previously vaccinated melanoma and ovarian cancer patients. Five of these (3 melanoma and 2 ovarian cancer patients) had received autologous GM-CSF-expressing whole tumor cell vaccines as an anticancer treatment. In all 5 patients antitumor immunity was induced, whereas 4 melanoma patients that were vaccinated with defined antigen vaccines did not show a clinical response. Tumor material obtained from the responding patients revealed dense infiltration of lymphocytes and neutrophils combined with tumor necrosis and vasculopathy. Although some skin toxicity was induced, the single treatment with MDX-CTLA4 (now called ipilimumab) was well tolerated.

It took another 7 years before it was demonstrated in a randomized controlled phase III trial, which compared ipilimumab plus a gp100 peptide vaccine to ipilimumab as a single agent and gp100 peptide vaccine as a single agent, that ipilimumab was capable of improving the median overall survival (OS) in metastatic melanoma patients that had received one prior systemic treatment for 4 months compared to gp100 vaccination. [11]. Based on this pivotal phase III trial, ipilimumab at a dose of 3 mg/kg for 4 consecutive 3-weekly infusions was approved for the treatment of metastatic melanoma in many countries around the world. In a second phase III trial, first-line dacarbazine, at that time the standard of care, was compared to the combination of dacarbazine plus ipilimumab (10 mg/kg) [12]. In this study the ipilimumab-containing arm also resulted in a statistically significant improvement in OS compared to dacarbazine alone. According to the data coming from phase II and III trials, ipilimumab induces objective clinical responses in 10–15% of patients. Most intriguing, however, is that in a substantial number of patients durable cancer regressions were observed. A pooled survival analysis based on close to 2,000 metastatic melanoma patients treated in phase II and III clinical trials revealed that at 3 years after the start of ipilimumab treatment 21% of patients were still alive, and that thereafter only a few patients succumbed to the disease [13].

CTLA-4 blockade in humans is accompanied by adverse events (AEs). Despite the fact that ipilimumab is a fully human IgG1 monoclonal antibody, infusion reactions do occur, although these are rare. Being a negative regulator of T cell immune responses and involved in the prevention of autoimmunity, blockade of CTLA-4 by ipilimumab may induce immune-related AEs (irAE) or AEs of special interest. These AEs include colitis, hepatitis, dermatitis, thyroiditis and hypophysitis, and closely resemble autoimmune diseases, but clearly lack the typical chronicity of diseases like inflammatory bowel disease, autoimmune hepatitis and dermatitis, and uveitis, etc. Analysis of AEs observed in a pivotal trial indicated that over 60% of patients develop irAE; however, in the vast majority these are mild [14]. In about 20% of treated patients grade 3/4 irAE were observed, oftentimes requiring hospitalization and high-dose corticosteroid therapy or other immunosuppressive drugs for resolution. Although management of these side effects require prompt intervention and a multidisciplinary approach, including gastroenterologists, endocrinologists and dermatologists, in experienced centers ipilimumab is a safe drug to administer.

Despite the knowledge that CTLA-4 is a key regulator of T cell immune responses, its exact mechanism of action in humans is still controversial. Recently, new insight was revealed on how checkpoint inhibition by CTLA-4 blockade changes the tumor-reactive T cell repertoire. First, in paired peripheral blood samples from metastatic melanoma or castrate refractory prostate cancer patients prior to treatment and following treatment with anti-CTLA-4, TCR Vβ CDR3 gene diversification and especially the persistence of clonotypes was strongly correlated with an improved outcome [15]. Whether these persistent clones contain melanoma-specific T cells is not known. Secondly, in another study a clear broadening of the melanoma-reactive T cell immune response upon treatment with ipilimumab was demonstrated. Here, HLA-A*0201 multimers harboring peptides derived from shared antigens belonging to the melanocyte differentiation antigens (MART-1, tyrosinase, gp100, etc.), cancer/testis gene products (MAGE family, LAGE, SSX-2, NY-eso-1, etc.) and overexpressed gene products (MELOE-1, GnTV, etc.) were utilized to follow changes in the frequency of peripheral blood CD8+ T cells against these antigens from before to after treatment [16]. Whereas the number of different melanoma-reactive T cell responses increased upon CTLA-4 blockade, a correlation with outcome could not me demonstrated. Third, ipilimumab treatment may expand T cell responses against epitopes derived from mutated gene products [17]. By exome DNA sequencing and RNA sequencing the total number of nonsynonymous somatic mutations in expressed genes can be quantified and potential mutated HLA-binding peptides be predicted. Using this information, CD8+ T cells specific for one mutated peptide or neoantigen was detected in a metastatic melanoma patient. Upon CTLA-4 blockade the frequency of the response increased 5-fold. Fourth, a clear correlation has been shown between the mutational load (the number of somatic mutations/megabase of DNA) of the tumor and the outcome of patients treated with ipilimumab [18]. These latter studies point in the direction of an important role for these neoantigens in relation to checkpoint inhibition.

In murine models, CTLA-4 inhibition by blocking antibodies is followed by depletion of CD4+/FoxP3+ regulatory T cells (Tregs) from the tumor microenvironment [19, 20]. As this results from an antibody-dependent cell-mediated cytotoxicity (ADCC) or from a complement-dependent cytotoxicity (CDC)-mediated process, the antibody subclass is of major importance. Ipilimumab is an IgG1 antibody that was picked for further clinical development because of a lack of ADCC/CDC activity in vitro. Tremelimumab, another anti-CTLA-4 monoclonal antibody in development, is a fully human IgG2 antibody also with no ADCC or CDC activity. Therefore, to date no strong evidence has been provided indicating towards the depletion of Tregs as a mechanism of CTLA-4 blockade in patients.

PD-1 and PD-L1

PD-1 was discovered in 1992 by Dr. Honjo and colleagues [21] as a gene encoding a protein belonging to the immunoglobulin superfamily, the expression of which was induced in dying thymocytes, hence its name programmed death receptor. It was later demonstrated that PD-1 expression was induced in lymphocytes and monocytes upon activation of the cells and was not involved in apoptosis. Mice deficient in PD-1 developed systemic lupus-like arthritis and glomerulonephritis late in life and, upon crossing with autoreactive 2C TCR transgenic mice, these animals developed a lethal graft-versus-host disease-like syndrome, with dense infiltration of inflammatory cells in dermis, lungs, kidneys and other tissues [22]. Depending on the mouse strain, PD-1-deficient animals could also develop dilated cardiomyopathy that was lethal in a substantial number of animals. As this chronic heart disease was absent in RAG-deficient animals, these results pointed towards an important role for PD-1 in maintaining immunological tolerance to self-antigens and prevention of autoimmune disease [23]. Shortly after this discovery, the genes encoding the ligands of PD-1 were cloned showing that these molecules belong to the B7 family [24, 25]. Although the tissue distribution of these ligands appears to be different, with a broader expression of PD ligand-1 (PD-L1; also called B7-H1) and a more restricted expression to the hematopoietic cell lineage of PD ligand-2 (PD-L2), engagement of PD-L1 or PD-L2 by PD-1-expressing T cells results in the downregulation of TCR-mediated signaling, culminating in a functional nonresponsiveness of the T cells. Expression of PD-L1 could also be induced on tumor cells, both in vivo in mouse tumor models as on human tumor cell lines. Expression of PD-L1 on tumor cells can result from intrinsic mechanisms, like PTEN deletion which activates the PI3-kinase pathway or from extrinsic stimuli like interferon-gamma secreted by surrounding activated T cells [26, 27]. This PD-L1 expression renders tumor-specific T cells inactive and, therefore, may serve as an immune escape mechanism in cancer immunotherapy. Indeed, blockade of the interaction between PD-L1-expressing tumor cells and PD-1-expressing tumor-specific T cells using PD-1 or PD-L1 blocking antibodies enhanced the cytolytic activity of the T cells [28–30]. Signaling through PD-L1 rendered tumor cells insensitive to T cell-induced apoptosis, suggesting that PD-L1 expression is a ubiquitous antiapoptosis molecule in cancer cells [31]. Expression of PD-L1 and/or PD-1 has been associated with patient outcome. In some tumors, expression of these molecules is correlated with poor survival, whereas in other tumors there is a clear association with improved outcome [see for instance 32–35]. In addition, tumor-infiltrating lymphocytes present in melanoma metastases often express high levels of PD-1, especially tumor-reactive cells [36, 37], clearly pointing towards the use of PD-1 expression as a selection marker for tumor-reactive T cells. In 2010, Brahmer et al. [38] demonstrated for the first time in a single infusion phase I study in patients with various tumor types (colorectal cancer, renal cell cancer, melanoma, non-small cell lung cancer, NSCLC, and castration refractory prostate cancer) clinical activity of MDX-1106, now named nivolumab, a fully human monoclonal antibody, at doses varying from 0.3 to 10 mg/kg. PD-1 receptor occupancy of >70% was reached for a period of at least 2 months following infusion irrespective of the dose level. The half-life of nivolumab was calculated to be 12–20 days. Furthermore, although irAE were observed in some patients, nivolumab appeared to have an acceptable safety profile. Based on these highly encouraging results, a series of clinical trials was initiated with nivolumab, pembrolizumab, another anti-PD-1 humanized antibody, and with MDX-1105 and MPDL3280A, two human anti-PD-L1 monoclonal antibodies.

Nivolumab was studied in a large multicohort phase I study with 296 patients but this time as repetitive dosing every 2 weeks (doses ranging from 0.1 to 10 mg/kg) in the same tumor types as the previous study [39]. Also in this study no dose-limiting toxicity was observed and efficacy could be demonstrated in metastatic melanoma (overall response rate, ORR 31%), metastatic renal cell cancer (ORR 27%) and NSCLC (ORR 18%). Grade 3–4 AEs were reported in only 14% of patients. Importantly, responses were durable. In the melanoma cohort of 107 patients, of whom 62% had had 2–5 prior systemic therapies, the median OS was over 16 months [40]. From phase I, nivolumab moved quickly into phase II and III trials. As the first-line treatment for metastatic melanoma patients not harboring the frequent BRAF V600 mutation, nivolumab was compared with dacarbazine, the standard of care chemotherapy [41]. In total 418 patients were randomized between the two arms. Importantly, at 1 year the OS was 72.9% for nivolumab compared to 42.1% for dacarbazine. At the time of the first reported analysis, the objective response rate for nivolumab-treated patients was 40%, of which 7.6% were complete responders (CRs), and for dacarbazine was 13.9% (1% CRs). Again, nivolumab was well tolerated with drug-related grade 3 or 4 AEs in less than 12% of patients, including immune-related AEs. Nivolumab has been approved by the US FDA for the treatment of metastatic melanoma.

Nivolumab has also been investigated in metastatic renal cell cancer (mRCC) and metastatic NSCLC (mNSCLC). In a randomized phase II trial comparing 3 doses of nivolumab in pretreated mRCC patients, an objective response rate of 20–22%, a median progression-free survival (PFS) of 4.0–4.2 months and OS of 24.7–25.5 months was demonstrated for the two highest dose arms (2 and 10 mg/kg, every 3 weeks) [42]. The OS is of particular interest, as this appears to be considerably longer in respect to historical data with other second- or third-line therapies (everolimus, axitinib, dovitinib). One has to await results from the phase III trials, which directly compare nivolumab to one of the approved second-line agents, everolimus. Also in mRCC patients, nivolumab had a very acceptable safety profile with 11% treatment-related grade 3 or 4 AEs.

Finally, in a large multicohort phase I trial [39], a total of 129 patients with heavily pretreated mNSCLC were included. Across the doses (1–10 mg/kg) tested the median OS was 9.9 months with 1-, 2- and 3-year survival of 42, 24 and 18%, respectively [43]. In 14% of patients treatment-related grade 3 or 4 AEs were reported, including three deaths due to a nivolumab-associated immune-related pneumonitis.

Similarly, phase I, II and III studies have been performed and are underway with pembrolizumab, a humanized engineered IgG4 monoclonal antibody blocking PD-1. Results from a large phase I study performed in patients with metastatic melanoma was reported at several time points [44, 45]. At first, superior activity appeared to be induced at a dose of 10 mg/kg every 2 weeks. However, in subsequent studies this difference could not be confirmed. A cohort of 173 patients with prior ipilimumab treatment was analyzed separately [45]. Two doses were compared, either 2 or 10 mg/kg every 3 weeks. In patients who failed ipilimumab treatment, pembrolizumab showed an objective response rate in 26%. Like nivolumab, pembrolizumab was well tolerated with grade 3–4 AEs in about 12–15% of the patients, of which fatigue was the most common. At 1 year, 58 and 63% of patients were alive, respectively. More recently, pembrolizumab was compared to ipilimumab as first-line treatment in a randomized controlled phase III trial. In total 834 metastatic melanoma patients either received pembrolizumab at 10 mg/kg every 2 or 3 weeks, or the standard schedule of ipilimumab. Objective response rates were reported to be around 33% for the pembrolizumab-treated patients and 12% for the ipilimumab-treated patients. The 6-month PFS rate was 47% for the anti-PD-1 arms and 26.5% for the anti-CTLA-4 arm, clearly illustrating a superior efficacy in terms of ORR and PFS for the PD-1-blocking antibody pembrolizumab. Pembrolizumab has also been approved by the US FDA for the treatment of metastatic melanoma.

Patients with cancers expressing PD-L1 prior to the initiation of treatment have a higher likelihood of responding to PD-1-blocking antibodies compared to patients with PD-L1-negative tumors, although the threshold for determination of PDL-1 expression is still being debated. Depending on the tumor type, this difference can be substantial. In melanoma patients, PD-L1-expressing tumors do better on anti-PD-1 therapy, but the difference with PD-L1-negative tumors is relatively small. Therefore, PD-L1 expression is not a biomarker for selecting patients who will respond to anti-PD-1 treatment. In NSCLC, the value of PD-L1 as a biomarker for response was investigated in a large phase I trial [46]. The ORR for all patients was 19.2% with a median PFS and OS of 3.7 months and 12 months, respectively. However, when patients were separated according to membranous PD-L1 expression, <1%, 1%< PD-L1 <49% and ≥50%, the ORR, PFS and OS were very similar for the first two groups, but substantially better for the third group. These results require prospective validation. Apart from PD-L1 expression by tumor cells, the mutational burden also has an impact on the outcome of anti-PD-1 treatment in NSCLC. It was already known that smokers have a higher number of somatic mutations than never smokers [47]. The correlation between mutational load and likelihood to respond to anti-CTLA-4 was clearly demonstrated for metastatic melanoma, and was caused by the induction of neoantigen-specific T cells. Recently, Rizvi et al. [48] showed a similar correlation between mutational burden and response to pembrolizumab for patients with metastatic NSCLC. In 1 responding patient a neoantigen-specific T cell response was induced and could be followed over time.

Despite the fact that anti-PD-1 and anti-PD-L1 antibodies have different effects – anti-PD-1 blocks interaction between PD-1 and PD-L1 and PD-L2, anti-PD-L1 blocks interaction between PD-L1 and PD-1 and B7.1 – the preclinical and clinical effects on tumor clearance by T cells appear very similar. The two currently most tested monoclonal antibodies, MDX-1105 and MPDL3280A, have shown activity in a variety of cancers. MDX-1105 was studied in a phase I trial with 207 patients with melanoma, NSCLC, RCC, colorectal cancer, pancreatic cancer, ovarian cancer and gastric cancer in several dose cohorts ranging from 0.3 to 10 mg/kg every 2 weeks. Objective clinical responses were seen in 6–17% of patients with NSCLC, melanoma, RCC and ovarian cancer. MDX-1105 was well tolerated and showed only a few grade 3 and 4 AEs (9%), typical for PD-1/PD-L1 checkpoint blockade. So far, MDX-1105 has not been further developed. MPDL3280A, a fully human engineered monoclonal antibody selected for absence of ADCC/CDC activity is being investigated in several tumor types. The first results were recently reported and showed activity over many tumor types (ORR 21%), with the highest rates in melanoma and NSCLC. Tumor tissue was available for biomarker research from many patients. In-depth analysis revealed that not only PD-L1 expression on tumor cells, but more importantly also PD-L1 expression on immune cells, like dendritic cells, macrophages and T cells within the tumor microenvironment prior to treatment, was associated with response to MPDL3280A [49]. Other potential biomarkers found in this study were CTLA-4 expression by immune cells and expression of the chemokine fractalkine. PD-L2 expression, however, was not associated with resistance to treatment.

Based on this phase I study, MPDL3280A is taken for further development given at a flat dose of 1,200 mg every 3 weeks. Within this large phase I study a cohort of 68 patients with metastatic urothelial bladder cancer was treated with MPDL3280A [50]. Based on the PD-L1 expression on infiltrating immune cells, classified in immunohistochemistry (IHC) 0, 1, 2 and 3, with IHC positivity defined as ≥5% of cells expressing PD-L1, objective responses were 43% for IHC 2/3 bladder cancer patients compared to 11% for the IHC 0/1 cancers. As most of the patients were heavily pretreated, including with platinum-based chemotherapy, induction of durable responses in these patients is quite an achievement and clearly illustrates the strength of immunotherapy by PD-1/PD-L1 blockade.

CTLA-4 Plus PD-1 Blockade

CTLA-4 and PD-1, although both being immune checkpoint molecules involved in maintaining peripheral tolerance and the prevention of autoimmunity, have distinct mechanisms and sites of action. CTLA-4 signaling occurs early during T cell activation, especially during T cell priming in the T cell-dependent areas of the lymph nodes, whereas PD-1 signaling occurs at effector sites in chronically inflamed tissue or tumors (fig. 1). Preclinical studies in a transplantable B16 melanoma model revealed the synergistic effect of dual checkpoint blockade. Upon combined inhibition of the inhibitory receptors CTLA-4 and PD-1, tumor-infiltrating T cell numbers increased, a change in the ratio of effector T cell to Tregs was induced, and effector T cell function was improved. Based on this and other studies, combination immunotherapy with ipilimumab and nivolumab was tested first in metastatic melanoma

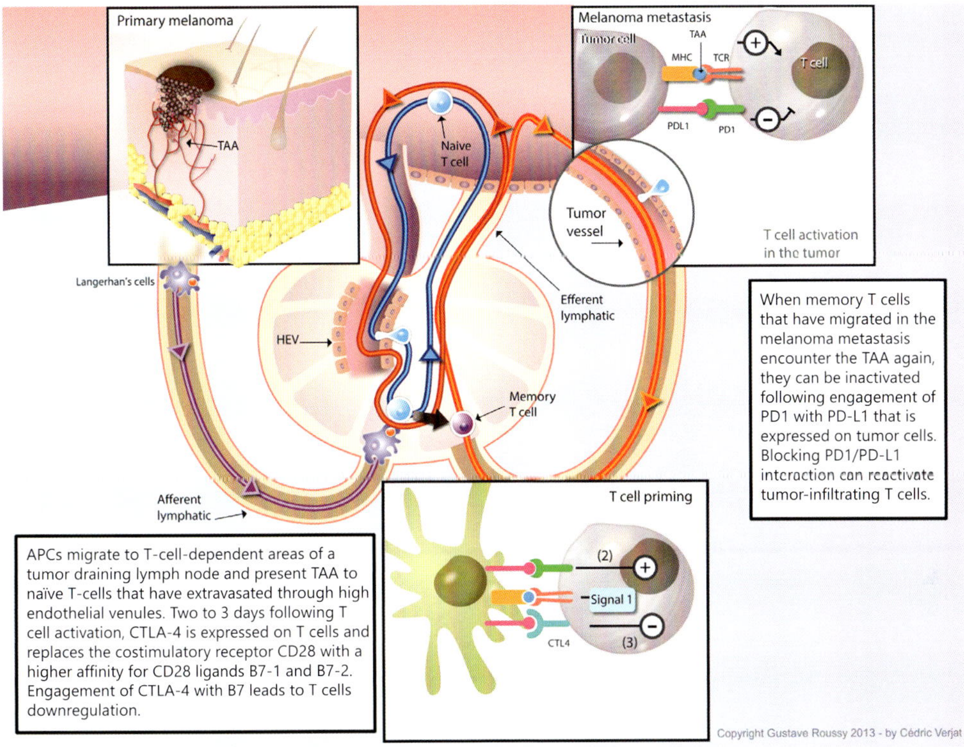

Fig. 1. Tumor specific T cells: patterns of migration. APCs = Antigen presenting cells; TAA = tumor associated antigens; HEV = high endothelial venules.

patients in a concurrent regimen. Wolchok et al. [51] reported on phase I in 53 patients treated with ipilimumab plus nivolumab in different dose cohorts. Objective clinical responses were found in 40% of patients. Most of these responses occurred early after initiation of the combination. Grade 3 and 4 toxicity were common (53% of patients), but were typical for immune checkpoint inhibition and manageable in the experienced centers where this trial was run. Recently, a randomized controlled phase II trial compared ipilimumab plus nivolumab to ipilimumab [52]. In BRAF wild-type melanoma patients (n = 72), objective clinical responses were observed in 61%, including 22% complete remissions, compared to 11% in ipilimumab without any CR. These results clearly demonstrate the superiority of combination checkpoint inhibition over ipilimumab, but one has to await the long-term beneficial effects in metastatic melanoma compared with inhibition of PD-1 alone. Meanwhile, this combination is being tested in other cancer patients, including mNSCLC and mRCC. Whether the results seen in melanoma patients will be matched in other cancer types remains to be seen. Similarly, the tolerability for this combination may be different in other tumor types as well.

Future Perspectives and Conclusion

Immune checkpoint inhibitors have set a paradigm shift in the treatment of metastatic cancer. Apart from CTLA-4 and the PD-1/PD-L1 axis, other immune checkpoints or inhibitory receptors have been discovered that, based on their cell surface expression, are targetable by monoclonal antibodies. TIM-3, LAG-3 and BTLA are examples of inhibitory receptors, some of which play a role in immunological tolerance, whereas the role of others appears much more subtle [53–55]. In preclinical models the combination of anti-TIM-3 plus anti-PD-1 or anti-LAG-3 and anti-PD-1 show synergism in tumor rejection, with an acceptable tolerance in animals so far [56, 57]. Therefore, apart from exploring the role of the already established molecules in other cancer types next to melanoma, NSCLC and RCC, new combinations will enter early clinical development as well. Furthermore, based on clinical responses in the various cancer types, checkpoint inhibitors as single agents or in combination will likely move towards earlier stages of disease and will be tested in adjuvant or neoadjuvant strategies, hoping to improve the OS of many cancer patients.

References

1 Brunet JF, Denizot F, Luciani MF, Roux-Dosseto M, Suzan M, Mattei MG, et al: A new member of the immunoglobulin superfamily – CTLA-4. Nature 1987; 328:267–270.

2 Linsley PS, Wallace PM, Johnson J, Gibson MG, Greene JL, Ledbetter JA, et al: Immunosuppression in vivo by a soluble form of the CTLA-4 T cell activation molecule. Science 1992;257:792–795.

3 Lenschow DJ, Zeng Y, Thistlethwaite JR, Montag A, Brady W, Gibson MG, et al: Long-term survival of xenogeneic pancreatic islet grafts induced by CTLA4Ig. Science 1992;257:789–792.

4 Chen L, Ashe S, Brady WA, Hellström I, Hellström KE, Ledbetter JA, et al: Costimulation of antitumor immunity by the B7 counterreceptor for the T lymphocyte molecules CD28 and CTLA-4. Cell 1992;71: 1093–1102.

5 Walunas TL, Lenschow DJ, Bakker CY, Linsley PS, Freeman GJ, Green JM, et al: CTLA-4 can function as a negative regulator of T cell activation. Immunity 1994;1:405–413.

6 Waterhouse P, Penninger JM, Timms E, Wakeham A, Shahinian A, Lee KP, et al: Lymphoproliferative disorders with early lethality in mice deficient in CTLA-4. Science 1995;270:985–988.

7 Leach DR, Krummel MF, Allison JP: Enhancement of antitumor immunity by CTLA-4 blockade. Science 1996;271:1734–1736.

8 van Elsas A, Hurwitz AA, Allison JP: Combination immunotherapy of B16 melanoma using anti-cytotoxic T lymphocyte-associated antigen 4 (CTLA-4) and granulocyte/macrophage colony-stimulating factor (GM-CSF)-producing vaccines induces rejection of subcutaneous and metastatic tumors accompanied by autoimmune depigmentation. J Exp Med 1999;190:355–366.

9 van Elsas A, Sutmuller RP, Hurwitz AA, Ziskin J, Villasenor J, Medema JP, et al: Elucidating the autoimmune and antitumor effector mechanisms of a treatment based on cytotoxic T lymphocyte antigen-4 blockade in combination with a B16 melanoma vaccine: comparison of prophylaxis and therapy. J Exp Med 2001;194:481–489.

10 Hodi FS, Mihm MC, Soiffer RJ, Haluska FG, Butler M, Seiden MV, et al: Biologic activity of cytotoxic T lymphocyte-associated antigen 4 antibody blockade in previously vaccinated metastatic melanoma and ovarian carcinoma patients. Proc Natl Acad Sci U S A 2003;100:4712–4717.

11 Hodi FS, O'Day SJ, McDermott DF, Weber RW, Haanen JB, Robert C, et al: Improved survival with ipilimumab in patients with metastatic melanoma. N Engl J Med 2010;363:711–723.

12 Robert C, Thomas L, Bondarenko I, O'Day S, Weber J, Garbe C, et al: Ipilimumab plus dacarbazine for previously untreated metastatic melanoma. N Engl J Med 2011;364:2517–2526.

13 Schadendorf D, Hodi FS, Robert C, Weber JS, Margolin K, Hamid O, et al: Pooled analysis of long-term survival data from phase II and phase III trials of ipilimumab in unresectable or metastatic melanoma. J Clin Oncol 2015;33:1889–1894.
14 Weber JS, Dummer R, Lebbe C, Hodi FS, et al: Patterns of onset and resolution of immune-related adverse events of special interest with ipilimumab. Cancer 2013;119:1675–1682.
15 Cha E, Klinger M, Hou Y, Cummings C, Ribas A, Faham M, et al: Improved survival with T cell clonotype stability after anti-CTLA-4 treatment in cancer patients. Sci Transl Med 2014;6:238ra70.
16 Kvistborg P, Philips D, Kelderman S, Hageman L, Ottensmeier C, Joseph-Pietras D, et al: Anti-CTLA-4 therapy broadens the melanoma-reactive CD8+ T cell response. Sci Transl Med 2014;6:254ra128.
17 van Rooij N, van Buuren MM, Philips D, Velds A, Toebes M, Heemskerk B, et al: Tumor exome analysis reveals neoantigen-specific T-cell reactivity in an ipilimumab-responsive melanoma. J Clin Oncol 2013;31:e439–e442.
18 Snyder A, Makarov V, Merghoub T, Yuan J, Zaretsky JM, Desrichard A, et al: Genetic basis for clinical response to CTLA-4 blockade in melanoma. N Engl J Med 2014;371:2189–2199.
19 Simpson TR, Li F, Montalvo-Ortiz W, Sepulveda MA, Bergerhoff K, Arce F, et al: Fc-dependent depletion of tumor-infiltrating regulatory T cells co-defines the efficacy of anti-CTLA-4 therapy against melanoma. J Exp Med 2013;210:1695–1710.
20 Selby MJ, Engelhardt JJ, Quigley M, Henning KA, Chen T, Srinivasan M, et al: Anti-CTLA-4 antibodies of IgG2a isotype enhance antitumor activity through reduction of intratumoral regulatory T cells. Cancer Immunol Res 2013;1:32–42.
21 Ishida Y, Agata Y, Shibahara K, Honjo T: Induced expression of PD-1, a novel member of the immunoglobulin gene superfamily, upon programmed cell death. EMBO J 1992;11:3887–3895.
22 Nishimura H, Nose M, Hiai H, Minato N, Honjo T: Development of lupus-like autoimmune diseases by disruption of the PD-1 gene encoding an ITIM motif-carrying immunoreceptor. Immunity 1999;11: 141–151.
23 Nishimura H, Okazaki T, Tanaka Y, Nakatani K, Hara M, Matsumori A, et al: Autoimmune dilated cardiomyopathy in PD-1 receptor-deficient mice. Science 2001;291:319–322.
24 Freeman GJ, Long AJ, Iwai Y, Bourque K, Chernova T, Nishimura H, et al: Engagement of the PD-1 immunoinhibitory receptor by a novel B7 family member leads to negative regulation of lymphocyte activation. J Exp Med 2000;192:1027–1034.
25 Latchman Y, Wood CR, Chernova T, Chaudhary D, Borde M, Chernova I, et al: PD-L2 is a second ligand for PD-1 and inhibits T cell activation. Nat Immunol 2001;2:261–268.
26 Parsa AT, Waldron JS, Panner A, Crane CA, Parney IF, Barry JJ, et al: Loss of tumor suppressor PTEN function increases B7-H1 expression and immunoresistance in glioma. Nat Med 2006;13:84–88.
27 Pardoll DM: The blockade of immune checkpoints in cancer immunotherapy. Nat Rev Cancer 2012;12: 252–264.
28 Iwai Y, Ishida M, Tanaka Y, Okazaki T, Honjo T, Minato N: Involvement of PD-L1 on tumor cells in the escape from host immune system and tumor immunotherapy by PD-L1 blockade. Proc Natl Acad Sci U S A 2002;99:12293–12297.
29 Blank C, Brown I, Peterson AC, Spiotto M, Iwai Y, Honjo T, et al: PD-L1/B7H-1 inhibits the effector phase of tumor rejection by T cell receptor (TCR) transgenic CD8+ T cells. Cancer Res 2004;64:1140–1145.
30 Hirano F, Kaneko K, Tamura H, Dong H, Wang S, Ichikawa M, et al: Blockade of B7-H1 and PD-1 by monoclonal antibodies potentiates cancer therapeutic immunity. Cancer Res 2005;65:1089–1096.
31 Azuma T, Yao S, Zhu G, Flies AS, Flies SJ, Chen L: B7-H1 is a ubiquitous antiapoptotic receptor on cancer cells. Blood 2008;111:3635–3643.
32 Thompson RH, Kuntz SM, Leibovich BC, Dong H, Lohse CM, Webster WS, et al: Tumor B7-H1 is associated with poor prognosis in renal cell carcinoma patients with long-term follow-up. Cancer Res 2006; 66:3381–3385.
33 Liu Y, Carlsson R, Ambjorn M, Hasan M, Badn W, Darabi A, et al: PD-L1 expression by neurons nearby tumors indicates better prognosis in glioblastoma patients. J Neurosci 2013;33:14231–14245.
34 Hamanishi J, Mandai M, Iwasaki M, Okazaki T, Tanaka Y, Yamaguchi K, et al: Programmed cell death 1 ligand 1 and tumor-infiltrating CD8+ T lymphocytes are prognostic factors of human ovarian cancer. Proc Natl Acad Sci U S A 2007;104:3360–3365.
35 Schalper KA, Velcheti V, Carvajal D, Wimberly H, Brown J, Pusztai L, et al: In situ tumor PD-L1 mRNA expression is associated with increased TILs and better outcome in breast carcinomas. Clin Canc Res 2014;20:2773–2782.
36 Ahmadzadeh M, Johnson LA, Heemskerk B, Wunderlich JR, Dudley ME, White DE, et al: Tumor antigen-specific CD8 T cells infiltrating the tumor express high levels of PD-1 and are functionally impaired. Blood 2009;114:1537–1544.
37 Gros A, Robbins PF, Yao X, Li YF, Turcotte S, Tran E, et al: PD-1 identifies the patient-specific CD8+ tumor-reactive repertoire infiltrating human tumors. J Clin Invest 2014;124:2246–2259.

38 Brahmer JR, Drake CG, Wollner I, Powderly JD, Picus J, Sharfman WH, et al: Phase I study of single-agent anti-programmed death-1 (MDX-1106) in refractory solid tumors: safety, clinical activity, pharmacodynamics, and immunologic correlates. J Clin Oncol 2010;28:3167–3175.
39 Topalian SL, Hodi FS, Brahmer JR, Gettinger SN, Smith DC, McDermott DF, et al: Safety, activity, and immune correlates of anti-PD-1 antibody in cancer. N Engl J Med 2012;366:2443–2454.
40 Topalian SL, Sznol M, McDermott DF, Kluger HM, Carvajal RD, Sharfman WH, et al: Survival, durable tumor remission, and long-term safety in patients with advanced melanoma receiving nivolumab. J Clin Oncol 2014;32:1020–1030.
41 Robert C, Long GV, Brady B, Dutriaux C, Maio M, Mortier L, et al: Nivolumab in previously untreated melanoma without BRAF mutation. N Engl J Med 2015;372:320–330.
42 Motzer RJ, Rini BI, McDermott DF, Redman BG, Kuzel TM, Harrison MR, et al: Nivolumab for metastatic renal cell carcinoma: results of a randomized phase II trial. J Clin Oncol 2015;33:1430–1437.
43 Gettinger SN, Horn L, Gandhi L, Spigel DR, Antonia SJ, Rizvi NA, et al: Overall survival and long-term safety of nivolumab (anti-programmed death 1 antibody, BMS-936558, ONO-4538) in patients with previously treated advanced non-small-cell lung cancer. J Clin Oncol 2015;33:2004–2012.
44 Hamid O, Robert C, Hodi FS, Hwu W-J, Wolchok JD, Joseph RW, et al: Safety and tumor responses with lambrolizumab (anti-PD-1) in melanoma. N Engl J Med 2013;369:134–144.
45 Robert C, Ribas A, Wolchok JD, Hodi FS, Hamid O, Kefford R, et al: Anti-programmed-death-receptor-1 treatment with pembrolizumab in ipilimumab-refractory advanced melanoma: a randomised dose-comparison cohort of a phase 1 trial. Lancet 2014; 384:1109–1117.
46 Garon EB, Rizvi NA, Hui R, Leighl N, Balmanoukian AS, Eder JP, et al: Pembrolizumab for the treatment of non-small-cell lung cancer. N Engl J Med 2015; 372:2018–2028.
47 Govindan R, Ding L, Griffith M, Subramanian J, Dees ND, Kanchi KL, et al: Genomic landscape of non-small cell lung cancer in smokers and never-smokers. Cell 2012;150:1121–1134.
48 Rizvi NA, Hellmann MD, Snyder A, Kvistborg P, Makarov V, Havel JJ, et al: Cancer immunology: mutational landscape determines sensitivity to PD-1 blockade in non-small cell lung cancer. Science 2015; 348:124–128.
49 Herbst RS, Soria J-C, Kowanetz M, Fine GD, Hamid O, Gordon MS, et al: Predictive correlates of response to the anti-PD-L1 antibody MPDL3280A in cancer patients. Nature 2015;515:563–567.
50 Powles T, Eder JP, Fine GD, Braiteh FS, Loriot Y, Cruz C, et al: MPDL3280A (anti-PD-L1) treatment leads to clinical activity in metastatic bladder cancer. Nature 2015;515:558–562.
51 Wolchok JD, Callahan MK, Postow MA, Lesokhin AM, Korman AJ, Wigginton JM, et al: Nivolumab plus ipilimumab in advanced melanoma. N Engl J Med 2013;369:122–133.
52 Postow MA, Chesney J, Pavlick AC, Robert C, Grossmann K, McDermott D, et al: Nivolumab and ipilimumab versus ipilimumab in untreated melanoma. N Engl J Med 2015;372:2006–2017.
53 Sabatos CA, Chakravarti S, Cha E, Schubart A, Sánchez-Fueyo A, Zheng XX, et al: Interaction of Tim-3 and Tim-3 ligand regulates T helper type 1 responses and induction of peripheral tolerance. Nat Immunol 2003;4:1102–1110.
54 Watanabe N, Gavrieli M, Sedy JR, Yang J, Fallarino F, Loftin SK, et al: BTLA is a lymphocyte inhibitory receptor with similarities to CTLA-4 and PD-1. Nat Immunol 2003;4:670–679.
55 Triebel F, Jitsukawa S, Baixeras E, Roman-Roman S, Genevee C, Viegas-Pequignot E, et al: LAG-3, a novel lymphocyte activation gene closely related to CD4. J Exp Med 1990;171:1393–1405.
56 Ngiow SF, von Scheidt B, Akiba H, Yagita H, Teng MWL, Smyth MJ: Anti-TIM3 antibody promotes T Cell IFN-γ-mediated antitumor immunity and suppresses established tumors. Cancer Res 2011;71: 3540–3551.
57 Woo SR, Turnis ME, Goldberg MV, Bankoti J, Selby M, Nirschl CJ, et al: Immune inhibitory molecules LAG-3 and PD-1 synergistically regulate T-cell function to promote tumoral immune escape. Cancer Res 2012;72:917–927.

Prof. John B.A.G. Haanen, MD, PhD
Divisions of Medical Oncology and Immunology
The Netherlands Cancer Institute, Plesmanlaan 121
1066 CX Amsterdam (The Netherlands)
E-Mail j.haanen@nki.nl

Michielin O, Coukos G (eds): Immuno-Oncology.
Prog Tumor Res. Basel, Karger, 2015, vol 42, pp 67–78 (DOI: 10.1159/000437185)

Radiotherapy and Immunotherapy: Improving Cancer Treatment through Synergy

Kobe Reynders · Dirk De Ruysscher

Department of Oncology, Experimental Radiation Oncology, KU Leuven – University of Leuven, Leuven, Belgium

Abstract

Radiotherapy is an important cornerstone in cancer treatment. Ionizing gamma-irradiation is capable of inducing DNA damage and consequential cell death in a precise and effective manner. In recent years it has become clear, however, that this is not the only relevant mechanism of action. Radiotherapy alters the immune composition of the tumor and influences upregulation of MHC I and cancer-testis antigens, inducing immunogenic cell death and supporting dendritic cell activation. Paradoxically, it also increases the relative ratio of regulatory T cells to CD4+ cells, which hampers an effective immune response. Nevertheless, the overall stimulating influence of irradiation on the immune system has been recognized and illustrated in preclinical studies as well as clinical case reports. There have been several attempts to use radiotherapy as an in situ vaccine. The basic rationale is a synergistic effect of different immune therapies like dendritic cell vaccination and CTLA-4 blockade with irradiation. Changes in the immune phenotype after radiotherapy can facilitate dendritic cell functioning. Immune therapy is also able to overcome the inhibitory pool of regulatory T cells through CTLA-4 inhibition, a weak point of radiotherapy. Although successful in preclinical models, there is still a lot of ground that needs to be covered. The optimal radiation dose is crucial, as well as timing and patient selection. Once these unknown parameters are explored, there is a lot of potential in the powerful combination of local immunization and systemic immune treatments for future novel cancer regimens.

For decades, cancer therapy has relied on three important pillars: surgery, chemotherapy and radiotherapy (RT). In the current curative setting, these are mostly applied in conjunction with one another. This multimodality approach tries to achieve a total eradication of tumor cells during the treatment process but lacks efficacious induction of long-term tumor resistance. Nevertheless, recurrent tumors and metastases still account for the major causes of tumor-associated death. Triggering a systemic and consistent immune response to malignant cells is vital to achieving a better and long-lasting tumor control. Unfortunately, efficient immune avoidance of tumor cells is one of the biggest hurdles in provoking this long-term tumor resistance. The absence of major histocompatibility complex I (MHC I), poor T cell influx to the tumor site, weak tumor-specific antigen expression and an anti-inflammatory environment are some of the most important factors that hinder an efficient cytotoxic immune response [1]. Targeted immune therapy often tackles only one of those domains. It explains why only moderate success is achieved with immune agents. Those that do provide an additional benefit are often accompanied by a variety of side effects and still struggle to handle larger tumor masses. For these reasons, it seems opportune that systemic immune agents have their place in multimodality treatments rather than monotherapy. It allows for synergistic interactions which enable a greater therapeutic benefit, dose reduction and potentially less systemic side effects. This appears to be crucial in a microenvironment where myeloid cells are severely altered towards a more immunosuppressive phenotype and immune evasion is pervasive [2].

The traditional application of local RT is based on inflicting DNA strand breaks through ionizing gamma-irradiation, which lead to necrosis and apoptosis. However, irradiation also has the capability of altering the immune phenotype of malignant cells in such a way that the threshold for triggering an immune response lowers. This way, RT provides an interesting addition to immune therapy as its effects go far beyond direct traditional cell kill. Irradiation also offers the advantages of being a local therapy. It dodges systemic side effects and effectively reduces tumor volume while still being able to modulate the immune context of the tumor.

Radiotherapy Increases Tumor Immunogenicity

On first sight, RT does not look like a natural ally for the immune system since irradiation destroys immune cells in the irradiated field [3]. This effect is, however, restricted in time and space. The long-term appeal of RT lies in the ability to counteract the immune evasive strategies of tumors. So how does irradiation achieve this? If irradiation wants to evoke an antitumor response outside of the treatment field, a local chain reaction has to be set in motion that can achieve a systemic answer to tumor proliferation. A cytokine and/or cytotoxic immune response seem to be the most obvious pathways. This implies that irradiation does not only achieve tumor control by inflicting DNA damage through ionizing beams, but also pushes an improved antitumor immune response. There are several factors that are important when assessing the immunogenic potential of RT, of which the most relevant ones are explained below. It is important to note that these should be evaluated in conjunction with one another.

Radiotherapy Changes the Cytokine Pattern

Recruiting and stimulating CD8+ T cells has proven to be essential for RT to combat tumor growth [4]. Dendritic cell activation is key to an efficient accrual of these CD8+ T cells. It is proposed that ablative irradiation leads to an increase in type I interferon, which is important for the cross-priming capacity of tumor-infiltrating dendritic cells [5, 6]. These findings suggest that the immune system and RT are more connected than might be initially suspected.

Radiotherapy Induces Immunogenic Cell Death

The concept that cell death is either tolerogenic, immunologically silent or even suppressive has been questioned for several years [7]. The theorem of a new immune-stimulating form of cell death is gaining ground quickly, stressing the importance of the way cell death is induced [8]. Initially this so-called immunogenic cell death (ICD) was observed only in certain chemotherapeutic agents, such as doxorubicin and oxaliplatin. Yet irradiation was also quickly profiled as a potent mediator [9–12].

The concept of ICD is based on the release of damage-associated molecular patterns (DAMPs) [13, 14]. These endogenous cellular molecules are released when endoplasmic reticulum stress and necrotic or apoptotic cell death occur, induced by certain treatment modalities like irradiation. This is a crucial step in the process of ICD. Different DAMPs have different roles, although they all aid in eliciting a long-term immune response. High-mobility group box proteins (HMGB) like HMGB1 and heat shock proteins (such as hsp70) act as proinflammatory signals, ATP is a find-me signal and calreticulin acts as an eat-me signal for antigen-presenting cells [14]. Unlike the freely released ATP, calreticulin and hsp70 are released as membrane blebs. The release of these danger signals leads to antigen engulfment and the stimulation of Toll-like receptors on dendritic cells (fig. 1). It has been suggested that autophagy and viral-like response signatures contribute to ICD as well, which illustrates that there is still a lot of unexplored terrain to be covered [15].

ICD facilitates tumor antigen presentation and enables an efficient activation of CD4+ and CD8+ T cells. For this reason it is believed that ICD is one of the most important contributors to a long-term antitumor response. The fact that irradiation is a local treatment with dose-dependent side effects provides it with additional benefits for inducing ICD when compared to systemic drugs.

Radiotherapy Changes Tumor Phenotype

Tumor development involves a change in the cellular protein pool. The surface presentation of these altered cellular peptides by MHC I is therefore vital in generating an adequate CD8+ T cell response. Downregulation of MHC I has already been mentioned as an important tool for tumors to escape immune surveillance. Irradiation is capable of augmenting the presence of MHC I on the tumor cell surface and accelerating protein degradation [16]. This leads to improved recognition of concealed malignant cells and, thus, a lower immunogenic threshold.

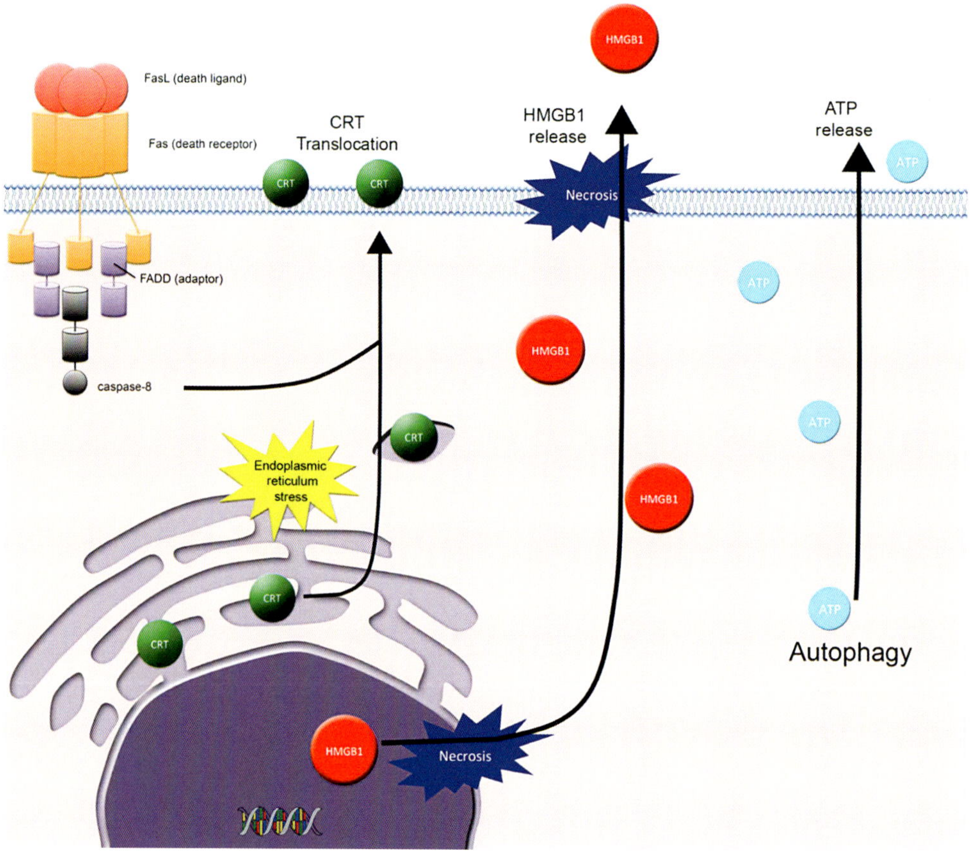

Fig. 1. The process of ICD. Irradiation causes endoplasmic reticulum stress, which lead to the translocation of calreticulin (CRT) to the cell surface. Upon cell death, both HMGB1 and ATP are released into the extracellular environment. From Golden et al. [56].

Further on, irradiation has the capability to increase de novo expression of cancer-testis antigens [17]. These tumor antigens do not appear on somatic tissues on normal cells. A heightened expression ensures a more distinct antigen profile on the tumor, avoiding tumors being recognized as 'self'. This holds true for tumor-associated antigens as well, which are massively released upon radiation-induced cell death [18]. However, not only the antigen presence is altered. Upregulation of Fas/CD95 and ICAM-I has been reported as well, increasing the susceptibility to effector T cells [19, 20].

Radiotherapy Alters the CD4+ T Cell Ratio and Function

In the context of tumor immunity, an important differentiation should be made between CD4+ CD25– T cells (Th) and CD4+ CD25+ FOX3+ T cells (regulatory T cells; Tregs). The latter are generally considered a negative prognostic factor since they inhibit a proper activation of cytotoxic T cells. They are also considered more

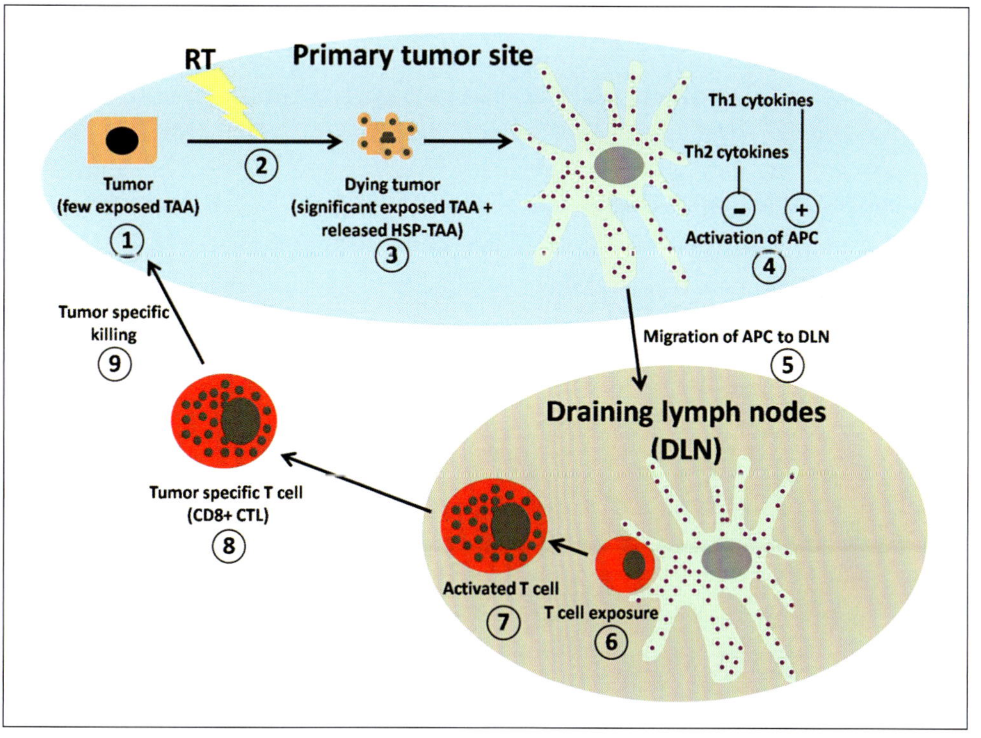

Fig. 2. Schematic representation of RT-induced antitumor immune regulation. **1** Within the primary tumor microenvironment (blue area) untreated tumors express few exposed tumor-associated antigens (TAAs). **2** Exposure to RT induces the dying tumors (**3**) to express significantly more TAAs on their surface and to release DAMPS (**4**), which are both taken up by professional APCs, resulting in their activation. Activation of APC is greatly enhanced (+) by the presence of Th1-type cytokines and significantly suppressed (–) by the presence of Th2-type cytokines. **5** Activated APCs migrate to draining lymph nodes (DLN; gray area). **6** Within the DLN, T cell exposure to APC is achieved by direct contact with activated APCs. **7** Activated T cells increase in size and granularity. **8** The activated T cells migrate from the DLN as tumor-specific T cells (CD8+ CTL) into the tumor microenvironment. **9** Within the tumor microenvironment CD8+ CTL perform tumor-specific killing. From Kaur and Asea [57].

radioresistant than Th cells. Consequently, the relative presence of Tregs goes up after irradiation since they are less prone to direct DNA damage [21, 22]. This is an example of an area in which irradiation might fall short. At the same time, it is important to recognize this because there are a lot of opportunities for complementary targeted agents in this field.

All these different mechanisms of action provide RT with a substantial toolset to fulfill an additional role aside from direct tumor cell kill. An overview of the immune regulation by irradiation is represented in figure 2. In this way irradiation is able to sculpt the tumor cell and microenvironment in such a manner that it helps to evoke an immune response rather than avoid it. It can be regarded as an in situ vaccination, a whole new mode of RT in cancer treatments.

The Abscopal Effect

The immunologic foundation of RT has also been indirectly illustrated in clinical practice. The 'abscopal effect' is a concept that originated in 1953, when Mole [23] used the term to describe tissue changes outside of the irradiated field. Since then, it has more commonly been used in case reports that describe a sudden regression of nonirradiated metastatic disease after palliative RT in only one area. Although rare, the abscopal effect has been defined in several malignancies over recent decades, including renal cell carcinoma [24, 25], hepatocellular carcinoma [26], malignant melanoma [27], lung adenocarcinoma [28] and several other pathologies. The difficulty lies in the retrospective nature of these case reports. Often the timing and method of imaging varies a lot from region to region which makes an appropriate comparison of these case reports difficult. There are also a lot of confounding factors to be wary of, such as concurrent or sequential systemic treatments, and spontaneous regression or a lack of clarity about the timing of events. At the same time, elements like sequential treatment or a perceived spontaneous regression might cover up true abscopal effects as well.

In clinical research, some authors have determined various cytokine levels in patients with a perceived abscopal effect. Ohba et al. [29] observed a rise of TNF-α after RT, whilst Nakanishi et al. [30] registered an increase in IL-18. These results have to be approached with caution since these cytokines are mostly arbitrarily chosen and no statistical correlation can be performed. They can be considered as a weak but interesting argument that there is a tight link between the RT-induced cytokine mixture and adaptive immunity.

The basics of the abscopal effect are difficult to unravel because of the rare events that occur in clinical practice. Preclinical data about the underlying mechanism of the abscopal effect is sparse but mostly attributes the abscopal effect to endogenous cytotoxic immune responses. Cytotoxic T cells have been established multiple times as an indispensable factor in achieving distant tumor control [31, 32]. Dendritic cells have proven to be fundamental as well [33, 34]. The interesting aspect here is that it appears to be very difficult to induce a consistent abscopal effect in vivo with RT alone, thus without concurrent immune treatments – why is this so?

Tumors suppress the immune system on two major levels. First and foremost, they reduce their own immunogenicity and resistance. For example, colon cancer cells evade cell death by secreting decoy receptors that bind and neutralize FasL [35]. T cell ignorance due to low tumor antigenicity and MHC I downregulation have already been mentioned. Secondly, malignant cells decrease the functionality and presence of innate and adaptive immune cells. Various tumors express death-inducing ligands that delete effector T cells [36]. Macrophages get pushed towards an immunosuppressive M2 phenotype through IL-4 and/or IL-13 [37]. Many other mechanisms are present, but clarifying them foregoes the goal of this chapter.

Recognizing this vast variety of immune evasion is important. As expected, irradiation does not tackle all these domains. RT is able to boost antigen variety and facilitate antigen presentation. However, it does not directly stimulate T cell functioning and numbers in an effective way. On the contrary, the relative pool of Tregs seems to rise after RT, thereby increasing the threshold for an operative immune response. Many variables account for this threshold, mostly determined by the tumor phenotype. Just like spontaneous regression, the abscopal effect is most likely limited to tumor phenotypes that possess fewer immunosuppressive mechanisms [38]. This explains why an abscopal effect is so exceptional.

Synergy with Immune Therapy

The diversity of tumor immune evasion is also key to understanding the potential synergy between RT and immune therapy. In general, they both share the common goal of initiating an adaptive immune response. Since irradiation is able to change the immune phenotype of the tumor, it forms an ideal addition to targeted immune agents and vaccination strategies.

In preclinical work the synergistic capabilities of RT have been displayed numerous times with various types of immune stimulation. Dendritic cells play a major role here. Fms-like tyrosine kinase receptor 3 ligand (Flt3-L), a growth factor for dendritic cells, has been combined with RT in multiple experiments. Chakravarty et al. [39] showed an improved disease-free survival and reduced nonirradiated pulmonary metastasis with this combination treatment. Monotherapy of either irradiation or Flt3-L failed to achieve the same results. Through depletion experiments, they established T cells as the vital components to achieving a systemic response. The same group also revealed a simultaneous increase in IFN-γ, IL-2 and IL-12 when using this combination treatment, indicating a procytotoxic cytokine pattern [40]. A more pure form of an abscopal effect was realized by Demaria et al. [33]. The combination of Flt3-L and RT on a breast tumor implanted on one flank led to a regression of a simultaneous tumor on the nonirradiated other flank, not observed in monotherapy. The importance of dendritic cells was illustrated once more by Teitz-Tennenbaum et al. [41]. They combined irradiation with a direct intratumoral injection of unpulsed dendritic cells in melanoma and sarcoma mouse models, which led to an improved overall survival and a reduced number of pulmonary metastases.

Other immune strategies yield the same results. IL-2 and RT led to a complete eradication of the primary tumor in a colon carcinoma mouse model while at the same time inhibiting liver metastases formation [42]. Preclinical success was also registered with ECI301, a macrophage inflammatory protein-1α variant. It is secreted by various leukocytes and recruits monocytes, dendritic cells, NK cells and T cells to create a more tumoricidal environment. It was used by Shiraishi et al. [31] in a fibrosarcoma model and delivered similar results when combined with RT as dendritic cells and IL-2.

In preclinical research, 9H10, a monoclonal antibody against cytotoxic lymphocyte antigen-4 (CTLA-4), has been used several times in conjunction with RT with excellent results on overall survival and tumor regression [32, 43]. CTLA-4 blockade is already a well-known treatment option in the clinical oncological landscape. Currently, it is the only mechanism that has demonstrated sound clinical results, especially for malignant melanoma. The sheer potential of this therapy has led to a substantial increase in experimental clinical usage across different pathologies and treatment schedules. There are already multiple case reports that indicate an abscopal effect after RT was given in conjunction with ipilimumab, a human monoclonal CTLA-4 antibody [44–47]. Both preclinical studies and case reports indicate a beneficial interaction. Theoretically, CTLA-4 inhibition should work well with RT since it counteracts the relative rise in Tregs after irradiation [48]. Based on this preclinical interaction, there have been multiple trials evaluating the combination treatment of ipilimumab and RT in castrate-resistant prostate cancer. In a phase I/II trial conducted by Slovin et al. [49], a palliative one-time dosage of 8 Gy was given right before the start of ipilimumab treatment. Ipilimumab showed a beneficial impact on prostate-specific antigen (PSA), but adding RT did not impact this trend significantly. A phase III trial with almost 800 randomized patients confirmed these results. A similar set-up of one-time 8 Gy followed by ipilimumab or placebo was used in castrate-resistant prostate cancer [50]. No benefit for overall survival was observed in the combination arm. After subgroup analysis, however, the researchers concluded that ipilimumab might be more beneficial in patients with more favorable prognostic features.

Another targeted agent, anti-PD-L1, has also achieved promising initial results with RT. Irradiation upregulates PD-L1 in the tumor microenvironment, which leads to a more immunosuppressive milieu. Adding anti-PD-L1 counteracts this negative influence by reducing tumor-infiltrating myeloid-derived suppressor cells and consequentially increasing the efficacy of irradiation [51].

Next to targeted agents, vaccination strategies form an important facet of immunotherapy as well. The beneficial effect of dendritic cell administration was clinically translated to a phase I trial in advanced hepatocellular carcinoma [52]. A vaccination cycle with immature dendritic cells was initiated 2 days after a single conformal RT dose of 8 Gy. Four out of 14 participants had minor responses and 2 had partial responses. An α-fetoprotein-specific immune response was established in 8 patients by cytokine release assay and in 7 patients by ELISPOT assay. In prostate cancer, a phase II trial was conducted that applied a PSA-encoded vaccine in conjunction with RT [53]. The vaccine was given together with granulocyte-macrophage colony-stimulating factor and low-dose IL-2. The classical RT guidelines in 2005 were applied (>70 Gy, 1.8- to 2-Gy fractions). This combination turned out to be safe and efficient in the generation of PSA-specific immune responses.

Many trials that combine RT with immune therapy are currently ongoing and many more will be initiated in the future [54]. Notably, there are many variables within RT that could influence its proimmunogenic capabilities which should be explored first. This might explain why clinical trials with ipilimumab have had various degrees of success.

Filling the Gaps

RT as a cancer treatment in daily practice is constantly evolving. Therapeutic regimens are highly dependent on imaging technology and the technique of dose delivery. The prescribed dose regimen varies from pathology to pathology and sometimes even from patient to patient. The total dose and dose per fraction are two of the most important parameters to influence tumor cell survival and normal tissue effects. It is only logical that they would influence the immunogenic effects of irradiation as well. Although preclinical studies have achieved good results with varying radiation doses, the RT regimen that is used receives very little attention at all. Both low (2–8 Gy) and high total doses (20–60 Gy) managed to show abscopal effects in preclinical studies and clinical case reports. Other in vitro studies detected a better activation of dendritic cells after normo- and hypofractionation, but not after a single dose of 15 Gy [55]. Secondly, the administration method of RT could matter as well, but no data is available. The impact of brachytherapy or image-guided RT, for example, on the immune phenotype of the tumor is not known at all. Different application methods could lead to immunogenic differences, which would impact the optimal combination treatment. Thirdly, the timing of RT and immune therapy is bound to be important. There are different theoretical benefits of administering RT before the immune treatment or the other way around. As discussed, RT leads to a time-restricted immune cell depletion, which might suggest that there should exist a window of opportunity between irradiation and immune therapy. Timing is likely to be dependent on the type of RT and immune therapy, and may thus change according to the treatment circumstances. Lastly, the rarity of the case reports illustrate an important aspect that should be explored further. Population selection will probably play a huge role in the potential success of radioimmunotherapy. Different tumors have a different immunogenicity and will yield different results. This also includes the disease stage. All these factors are important for RT and immune therapy and they require further investigation to achieve consistent clinical success.

Conclusion

RT is an important instrument for cancer treatment. The efficient cell kill it causes and the possibility to conduct a very precise localized treatment make it ideal for combatting tumor growth. At the same time, these features inhibit it from acting as a systemic treatment. Preclinical studies have shown that irradiation is also able to inflict an important level of phenotype modification in regards to tumor immunogenicity. This has been illustrated by several clinical case reports as well. Although this only rarely leads to a systemic immune response by itself, it does open up a lot of potential for combination treatments. Immune agents and vaccination strategies are the most suitable candidates with which to achieve a synergistic treatment effect. Early clinical trials are promising, but also illustrate the variable outcome. Therefore, further research is needed to determine optimal radiation dosages, timing of both treatments and patient selection. This combination regimen holds the potential to greatly improve patient care, but there are still a lot of unknown parameters to be discovered.

References

1 Zitvogel L, Tesniere A, Kroemer G: Cancer despite immunosurveillance: immunoselection and immunosubversion. Nat Rev Immunol 2006;6:715–727.

2 Gabrilovich DI, Ostrand-Rosenberg S, Bronte V: Coordinated regulation of myeloid cells by tumours. Nat Rev Immunol 2012;12:253–268.

3 Anderson RE, Warner NL: Ionizing radiation and the immune response. Adv Immunol 1976;24:215–335.

4 Gupta A, Probst HC, Vuong V, Landshammer A, Muth S, Yagita H, et al: Radiotherapy promotes tumor-specific effector CD8+ T cells via dendritic cell activation. J Immunol 2012;189:558–566.

5 Burnette BC, Liang H, Lee Y, Chlewicki L, Khodarev NN, Weichselbaum RR, et al: The efficacy of radiotherapy relies upon induction of type I interferon-dependent innate and adaptive immunity. Cancer Res 2011;71:2488–2496.

6 Lim JY, Gerber SA, Murphy SP, Lord EM: Type I interferons induced by radiation therapy mediate recruitment and effector function of $CD8^+$ T cells. Cancer Immunol Immunother 2014;63:259–271.

7 Voll RE, Herrmann M, Roth EA, Stach C, Kalden JR, Girkontaite I: Immunosuppressive effects of apoptotic cells. Nature 1997;390:350–351.

8 Casares N, Pequignot MO, Tesniere A, Ghiringhelli F, Roux S, Chaput N, et al: Caspase-dependent immunogenicity of doxorubicin-induced tumor cell death. J Exp Med 2005;202:1691–1701.

9 Apetoh L, Ghiringhelli F, Tesniere A, Obeid M, Ortiz C, Criollo A, et al: Toll-like receptor 4-dependent contribution of the immune system to anticancer chemotherapy and radiotherapy. Nat Med 2007;13:1050–1059.

10 Obeid M, Panaretakis T, Joza N, Tufi R, Tesniere A, van Endert P, et al: Calreticulin exposure is required for the immunogenicity of γ-irradiation and UVC light-induced apoptosis. Cell Death Differ 2007;14:1848–1850.

11 Gameiro SR, Jammeh ML, Wattenberg MM, Tsang KY, Ferrone S, Hodge JW: Radiation-induced immunogenic modulation of tumor enhances antigen processing and calreticulin exposure, resulting in enhanced T-cell killing. Oncotarget 2014;5:403–416.

12 Perez CA, Fu A, Onishko H, Hallahan DE, Geng L: Radiation induces an antitumour immune response to mouse melanoma. Int J Radiat Biol 2009;85:1126–1136.

13 Kroemer G, Galluzzi L, Kepp O, Zitvogel L: Immunogenic cell death in cancer therapy. Annu Rev Immunol 2013;31:51–72.

14 Krysko DV, Garg AD, Kaczmarek A, Krysko O, Agostinis P, Vandenabeele P: Immunogenic cell death and DAMPs in cancer therapy. Nat Rev Cancer 2012;12:860–875.

15 Garg AD, Martin S, Golab J, Agostinis P: Danger signalling during cancer cell death: origins, plasticity and regulation. Cell Death Differ 2014;21:26–38.

16 Reits EA, Hodge JW, Herberts CA, Groothuis TA, Chakraborty M, Wansley EK, et al: Radiation modulates the peptide repertoire, enhances MHC class I expression, and induces successful antitumor immunotherapy. J Exp Med 2006;203:1259–1271.

17 Sharma A, Bode B, Wenger RH, Lehmann K, Sartori AA, Moch H, et al: γ-Radiation promotes immunological recognition of cancer cells through increased expression of cancer-testis antigens in vitro and in vivo. PLoS One 2011;6:e28217.

18 Hannani D, Sistigu A, Kepp O, Galluzzi L, Kroemer G, Zitvogel L: Prerequisites for the antitumor vaccine-like effect of chemotherapy and radiotherapy. Cancer J 2011;17:351–358.

19 Garnett CT, Palena C, Chakraborty M, Tsang KY, Schlom J, Hodge JW: Sublethal irradiation of human tumor cells modulates phenotype resulting in enhanced killing by cytotoxic T lymphocytes. Cancer Res 2004;64:7985–7994.

20 Chakraborty M, Abrams SI, Camphausen K, Liu K, Scott T, Coleman CN, et al: Irradiation of tumor cells up-regulates Fas and enhances CTL lytic activity and CTL adoptive immunotherapy. J Immunol 2003; 170:6338–6347.
21 Kachikwu EL, Iwamoto KS, Liao YP, DeMarco JJ, Agazaryan N, Economou JS, et al: Radiation enhances regulatory T cell representation. Int J Radiat Oncol Biol Phys 2011;81:1128–1135.
22 Balogh A, Persa E, Bogdandi EN, Benedek A, Hegyesi H, Safrany G, et al: The effect of ionizing radiation on the homeostasis and functional integrity of murine splenic regulatory T cells. Inflamm Res 2013; 62:201–212.
23 Mole R: Whole body irradiation therapy – radiobiology or medicine? Br J Radiol 1953;26:234–241.
24 Fairlamb DJ: Spontaneous regression of metastases of renal cancer: a report of two cases including the first recorded regression following irradiation of a dominant metastasis and review of the world literature. Cancer 1981;47:2102–2106.
25 MacManus MP, Harte RJ, Stranex S: Spontaneous regression of metastatic renal cell carcinoma following palliative irradiation of the primary tumour. Ir J Med Sci 1994;163:461–463.
26 Okuma K, Yamashita H, Niibe Y, Hayakawa K, Nakagawa K: Abscopal effect of radiation on lung metastases of hepatocellular carcinoma: a case report. J Med Case Rep 2011;5:111.
27 Kingsley DP: An interesting case of possible abscopal effect in malignant melanoma. Br J Radiol 1975;48: 863–866.
28 Antoniades J, Brady LW, Lightfoot DA: Lymphangiographic demonstration of the abscopal effect in patients with malignant lymphomas. Int J Radiat Oncol Biol Phys 1977;2:141–147.
29 Ohba K, Omagari K, Nakamura T, Ikuno N, Saeki S, Matsuo I, et al: Abscopal regression of hepatocellular carcinoma after radiotherapy for bone metastasis. Gut 1998;43:575–577.
30 Nakanishi M, Chuma M, Hige S, Asaka M: Abscopal effect on hepatocellular carcinoma. Am J Gastroenterol 2008;103:1320–1321.
31 Shiraishi K, Ishiwata Y, Nakagawa K, Yokochi S, Taruki C, Akuta T, et al: Enhancement of antitumor radiation efficacy and consistent induction of the abscopal effect in mice by ECI301, an active variant of macrophage inflammatory protein-1α. Clin Cancer Res 2008;14:1159–1166.
32 Demaria S, Kawashima N, Yang AM, Devitt ML, Babb JS, Allison JP, et al: Immune-mediated inhibition of metastases after treatment with local radiation and CTLA-4 blockade in a mouse model of breast cancer. Clin Cancer Res 2005;11:728–734.
33 Demaria S, Ng B, Devitt ML, Babb JS, Kawashima N, Liebes L, et al: Ionizing radiation inhibition of distant untreated tumors (abscopal effect) is immune mediated. Int J Radiat Oncol Biol Phys 2004;58:862–870.
34 Akutsu Y, Matsubara H, Urashima T, Komatsu A, Sakata H, Nishimori T, et al: Combination of direct intratumoral administration of dendritic cells and irradiation induces strong systemic antitumor effect mediated by GRP94/gp96 against squamous cell carcinoma in mice. Int J Oncol 2007;31:509–515.
35 Pitti RM, Marsters SA, Lawrence DA, Roy M, Kischkel FC, Dowd P, et al: Genomic amplification of a decoy receptor for Fas ligand in lung and colon cancer. Nature 1998;396:699–703.
36 Gastman BR, Atarshi Y, Reichert TE, Saito T, Balkir L, Rabinowich H, et al: Fas ligand is expressed on human squamous cell carcinomas of the head and neck, and it promotes apoptosis of T lymphocytes. Cancer Res 1999;59:5356–5364.
37 Gordon S, Martinez FO: Alternative activation of macrophages: mechanism and functions. Immunity 2010;32:593–604.
38 Ricci SB, Cerchiari U: Spontaneous regression of malignant tumors: importance of the immune system and other factors (review). Oncol Lett 2010;1: 941–945.
39 Chakravarty PK, Alfieri A, Thomas EK, Beri V, Tanaka KE, Vikram B, et al: Flt3-ligand administration after radiation therapy prolongs survival in a murine model of metastatic lung cancer. Cancer Res 1999;59:6028–6032.
40 Chakravarty PK, Guha C, Alfieri A, Beri V, Niazova Z, Deb NJ, et al: Flt3L therapy following localized tumor irradiation generates long-term protective immune response in metastatic lung cancer: its implication in designing a vaccination strategy. Oncology 2006;70:245–254.
41 Teitz-Tennenbaum S, Li Q, Rynkiewicz S, Ito F, Davis MA, McGinn CJ, et al: Radiotherapy potentiates the therapeutic efficacy of intratumoral dendritic cell administration. Cancer Res 2003;63:8466–8475.
42 Yasuda K, Nirei T, Tsuno NH, Nagawa H, Kitayama J: Intratumoral injection of interleukin-2 augments the local and abscopal effects of radiotherapy in murine rectal cancer. Cancer Sci 2011;102:1257–1263.
43 Dewan MZ, Galloway AE, Kawashima N, Dewyngaert JK, Babb JS, Formenti SC, et al: Fractionated but not single-dose radiotherapy induces an immune-mediated abscopal effect when combined with anti-CTLA-4 antibody. Clin Cancer Res 2009; 15:5379–5388.
44 Golden EB, Demaria S, Schiff PB, Chachoua A, Formenti SC: An abscopal response to radiation and ipilimumab in a patient with metastatic non-small cell lung cancer. Cancer Immunol Res 2013;1:365–372.

45 Hiniker SM, Chen DS, Reddy S, Chang DT, Jones JC, Mollick JA, et al: A systemic complete response of metastatic melanoma to local radiation and immunotherapy. Transl Oncol 2012;5:404–407.
46 Postow MA, Callahan MK, Barker CA, Yamada Y, Yuan J, Kitano S, et al: Immunologic correlates of the abscopal effect in a patient with melanoma. N Engl J Med 2012;366:925–931.
47 Stamell EF, Wolchok JD, Gnjatic S, Lee NY, Brownell I: The abscopal effect associated with a systemic anti-melanoma immune response. Int J Radiat Oncol Biol Phys 2013;85:293–295.
48 Baba J, Watanabe S, Saida Y, Tanaka T, Miyabayashi T, Koshio J, et al: Depletion of radio-resistant regulatory T cells enhances antitumor immunity during recovery from lymphopenia. Blood 2012;120:2417–2427.
49 Slovin SF, Higano CS, Hamid O, Tejwani S, Harzstark A, Alumkal JJ, et al: Ipilimumab alone or in combination with radiotherapy in metastatic castration-resistant prostate cancer: results from an open-label, multicenter phase I/II study. Ann Oncol 2013; 24:1813–1821.
50 Kwon ED, Drake CG, Scher HI, Fizazi K, Bossi A, van den Eertwegh AJ, et al: Ipilimumab versus placebo after radiotherapy in patients with metastatic castration-resistant prostate cancer that had progressed after docetaxel chemotherapy (CA184-043): a multicentre, randomised, double-blind, phase 3 trial. Lancet Oncol 2014;15:700–712.
51 Deng L, Liang H, Burnette B, Beckett M, Darga T, Weichselbaum RR, et al: Irradiation and anti-PD-L1 treatment synergistically promote antitumor immunity in mice. J Clin Invest 2014;124:687–695.
52 Chi KH, Liu SJ, Li CP, Kuo HP, Wang YS, Chao Y, et al: Combination of conformal radiotherapy and intratumoral injection of adoptive dendritic cell immunotherapy in refractory hepatoma. J Immunother 2005;28:129–135.
53 Gulley JL, Arlen PM, Bastian A, Morin S, Marte J, Beetham P, et al: Combining a recombinant cancer vaccine with standard definitive radiotherapy in patients with localized prostate cancer. Clin Cancer Res 2005;11:3353–3362.
54 Vacchelli E, Vitale I, Tartour E, Eggermont A, Sautes-Fridman C, Galon J, et al: Trial watch: anticancer radioimmunotherapy. Oncoimmunology 2013;2: e25595.
55 Kulzer L, Rubner Y, Deloch L, Allgauer A, Frey B, Fietkau R, et al: Norm- and hypo-fractionated radiotherapy is capable of activating human dendritic cells. J Immunotoxicol 2014;11:328–336.
56 Golden EB, Pellicciotta I, Demaria S, Barcellos-Hoff MH, Formenti SC: The convergence of radiation and immunogenic cell death signaling pathways. Front Oncol 2012;2:88.
57 Kaur P, Asea A: Radiation-induced effects and the immune system in cancer. Front Oncol 2012;2:191.

Kobe Reynders, MD
KU Leuven, Experimental Radiation Oncology
Herestraat 49, bus 815
BE–3000 Leuven (Belgium)
E-Mail kobe.reynders@med.kuleuven.be

Michielin O, Coukos G (eds): Immuno-Oncology.
Prog Tumor Res. Basel, Karger, 2015, vol 42, pp 79–94 (DOI: 10.1159/000437186)

Combinations Therapies

Niels Reinmuth • Martin Reck

Department of Thoracic Oncology, LungenClinic Großhansdorf, Grosshansdorf and Airway Research Center North, German Center for Lung Research (DZL), Heidelberg, Germany

Abstract

Immunotherapy of cancer encompasses different strategies that elicit or enhance the immune response against tumors. The first results from clinical studies have provided promising data for the treatment of lung cancer patients with immunomodulating monotherapies. To improve the potential benefit of cancer immunotherapy, synergistic combinations of the various immunotherapy approaches or of different elements within each of the immunotherapy approaches are being explored. The rationale typically involves different but complementary mechanisms of action, eventually impinging on more than one immune system mechanism. As a prominent example, the simultaneous blockade of PD-1 and CTLA-4 is giving rise to therapeutic synergy, while still offering room for efficacy improvement. Moreover, combinations of immunomodulating agents with chemotherapy or targeted molecules are being tested. Animal models suggest that immunotherapies in combination with these various options offer evidence for synergistic effects and are likely to radically change cancer treatment paradigms. However, data obtained so far indicate that toxic side effects are also potentiated, which may even restrict the selection of patients that are suitable for these combinational approaches. Advancing the field of combinatorial immunotherapy will require changes in the way investigational agents are clinically developed as well as novel experimental end-points for efficacy evaluation. However, this combined therapeutic manipulation of both tumor and stromal cells may lead to a dramatic change in the therapeutic options of lung cancer patients in any disease stage that can only grossly be appreciated by the current studies.

Cancer evolves to exploit multiple mechanisms in order to avoid immune cell recognition and elicit an innate or adaptive immune response. These mechanisms include downregulation of MHC expression, rendering tumor cells insensitive to T cells, and the secretion of factors that suppress T cell responsiveness [1]. Tumors may also recruit or convert inflammatory cells, such as regulatory T cells, and myeloid-derived suppressor cells further suppress T cell responses [2]. Moreover, multiple interactions with the tumor microenvironment, such as stimulating the release of angiogenic factors, may also contribute to the inefficient presentation of tumor antigens and escaping cancer immunosurveillance [3].

In turn, immunotherapy aims to target and harness the patient's own immune system by manipulating the unique properties of the immune system of specificity, adaptability and memory. This strategy includes activating and enhancing innate and adaptive immune effector mechanisms that may be achieved in multiple ways, such as vaccination with tumor antigens, treatment with cytokines and the blockade of key immune checkpoints. Hence, this approach is very different from 'traditional' treatment approaches which target the tumor cells or tumor blood supply. Due to the complex regulation of immunosurveillance various therapeutic approaches are being explored in preclinical and clinical trials. As an example, because some tumors engage checkpoint pathways to escape antitumor immune responses, checkpoint molecule blockade is being pursued as a therapeutic anticancer strategy. Despite promising results in clinical trials in lung cancer and other tumor types demonstrating long-term disease control with moderate toxicity for at least a subset of patients, some cancers fail immune interventions. The secretion of immunosuppressive factors, defective antigen presentation, and expression of death ligands or serpins have been regarded as possible escape mechanisms [4]. As proved successful in many other examples of advances in the history of medicine, combining different strategies may result in an improved response and favorable outcome. Currently, some approaches are being investigated in clinical trials in lung cancer patients. However, very limited data from these trials have been reported so far, which renders firm conclusions difficult. Still, important aspects of the rationales and clinical experiences of combination immunotherapies are described in the following discussion.

Combination of Immunotherapy with Chemotherapy

Due to the potent immunosuppressive effect usually associated with chemotherapy, the use of conventional cancer chemotherapy in combination with cancer immunotherapy has been questioned. However, both preclinical and early clinical studies have seen no detrimental effects in immune responses to vaccines in patients when a vaccine is given in combination with certain chemotherapeutic agents [5]. In contrast, some cytotoxic agents such as anthracyclines have been associated with increased primary cell-mediated immunity [6]. Moreover, both preclinical and clinical findings have suggested that conventional chemotherapy and radiation therapy can induce immune responses against antigens generated in dying tumor cells and the interaction of tumor-cell secreted (high-mobility group box 1; HMGB1) alarmin proteins on Toll-like receptor 4 (TLR4) expressed by dendritic cells [7]. In addition, patients with nonmetastatic breast cancer but locoregional lymph node involvement and carrying a polymorphism in TLR4 that results in reduced binding to HMGB1 did experience a relapse of their disease following adjuvant anthracyclines therapy after local surgery significantly earlier compared to patients without the mutation [7]. Also, chemotherapy may foster the release of tumor-specific antigens from dying tumor cells and improve the immunogenicity of the microenvironment [8]. Finally, in a series of mice models, treatment with various cytotoxic agents, such as paclitaxel and cisplatin, made tumor cells more susceptible to the cytotoxic effect of T lymphocytes and induced apoptosis in a broader range of tumor cells [9].

The recent development of immunomodulating agents in lung cancer is led by the generation of monoclonal antibodies that target specific components of the immune system. Although several approaches are possible, those agents that target inhibitors at checkpoints of the immune response activation cascade are currently leading the way and have the most advanced clinical development programs. For example, cytotoxic T lymphocyte-associated antigen 4 (CTLA-4) is an immune checkpoint receptor that is expressed exclusively on T cells where it primarily regulates the amplitude of the early stages of T cell activation [4]. Tumor-mediated activation of the CTLA-4 inhibitory pathway decreases the amplitude of the T cell response [10]. Hence, strategies of blocking the activation of the CTLA-4 pathway with anti-CTLA-4 monoclonal antibodies to increase antitumor immune activity have been explored in preclinical tumor models [10, 11]. Blocking the interaction of CTLA-4 with its ligands (CD80/CD86) augmented T cell activation and proliferation, leading to tumor infiltration by T cells and tumor regression [12]. However, despite its preclinical efficacy and promising clinical activity against late-stage metastatic melanoma, the exact mechanism by which antibody targeting of CTLA-4 leads to tumor cell death remains unclear [4]. Early clinical trials with humanized CTLA-4 antibodies as monotherapy in patients with advanced disease demonstrated objective clinical responses in about 10% of patients with melanoma [13]. Because the mechanism of action of CTLA-4 antibodies is not dependent on tumor cells expressing a particular antigenic target, it was reasoned that it may also have activity against multiple tumor types.

Therefore, clinical trials have been established to evaluate the efficacy of anti-CTLA-4 therapy in lung cancer employing specific monoclonal antibodies. In studies on patients with advanced non-small cell lung cancer (NSCLC), the anti-CTLA-4 monoclonal antibody ipilimumab has been paired with chemotherapy, specifically a doublet of paclitaxel and carboplatin. In a phase II study, treatment with chemotherapy (paclitaxel and carboplatin) plus a phased application of the anti-CTLA-4 monoclonal antibody ipilimumab significantly improved immune-related progression-free survival (irPFS) in previously untreated patients with advanced NSCLC when compared with chemotherapy alone (5.7 vs. 4.6 months; hazard ratio = 0.72, p = 0.05) [14]. Interestingly, concurrent application of ipilimumab with chemotherapy did not provide the same level of clinical benefit. Similar results have been reported from studies focusing on small cell lung cancer (SCLC). In patients with extensive disease SCLC a significant improvement in irPFS was achieved using a phased schedule of ipilimumab with carboplatin/paclitaxel treatment (6.4 vs. 5.3 months; hazard ratio = 0.64, p = 0.03) but not with concurrent treatment (hazard ratio = 0.75, p = 0.11) [15]. Ongoing phase III studies are currently evaluating a phased addition of ipilimumab to combination chemotherapy in patients with advanced, recurrent squamous NSCLC and extensive stage disease SCLC, respectively. Another anti-CTLA-4 antibody, tremelimumab, is also in clinical development for NSCLC in various combinations. An open-label, phase II clinical trial did not demonstrate superiority of tremelimumab over the best supportive care alone as a maintenance therapy in patients with locally advanced or metastatic NSCLC and after treatment with a first-line platinum-based therapy [16]. However, an objective response rate of 4.8% seen only in the investigational arm warrants further investigation.

Moreover, the anti-PD-1 antibody nivolumab, another checkpoint inhibitor targeting the programmed death-1 (PD-1) receptor, is also being investigated in combination with chemotherapy in NSCLC. In an updated interim analysis of a phase I study, nivolumab plus platinum-based chemotherapy in stage IIIB/IV NSCLC yielded objective response rates of 33–50% and a PFS at 24 weeks ranging from 36 to 71% [17]. Interestingly, objective responses were observed in patients with both squamous and non-squamous NSCLC. Moreover, encouraging 1-year overall survival rates (59–87%) across all cohorts have been reported [17].

Combination of Various Immunotherapies

Due to the fact that the immune response is tightly regulated by a host of humoral and cellular interactions, it seems logical to further enhance the magnitude of the antitumor immune effect by combining or sequencing immunotherapies that target distinct immune pathways. This is especially important considering the fact that tumors may often use more than one mechanism to avoid immune detection. Several immunotherapeutic agents, including a range of vaccines and immuno-oncology agents, are currently in development and demonstrate clinical efficacy. As reviewed elsewhere, most clinical trials have evaluated immunomodulatory agents as monotherapies showing substantial promise [18]. For example, immune checkpoints refer to a broad range of inhibitory pathways in the immune system that are crucial for maintaining self-tolerance and preventing excessive, prolonged and potentially deleterious T cell activity in peripheral tissues [4]. There is growing evidence that lung cancer can utilize these immune checkpoints to circumvent the antitumor immune response.

A variety of inhibitors are in clinical development that target different immune checkpoints, such as the CTLA-4 receptor, the PD-1 receptor or its ligands, the killer cell immunoglobulin-like receptor (KIR) and lymphocyte-activation gene-3 (LAG-3)

[18]. Moreover, several data support the combination of immunomodulating agents that target different immune escape mechanisms. For example, there is a strong rationale for dual T cell checkpoint inhibition by blocking both CTLA-4 and PD-1. While CTLA-4 is expressed exclusively on T cells and contributes to T cell activation upon recognition of MHC-antigen complexes, the PD-1 receptor is expressed on T cells, B cells and natural killer cells, and may also affect these cell types [10, 19]. Furthermore, while agents targeting CTLA-4 act in the periphery and affect the early stages of cytotoxic T lymphocyte activation, those targeting the PD-1 pathway act more centrally within the tumor microenvironment itself, and can impact on the late stages of T cell activation [4, 10].

In a recent phase I study of 53 patients with advanced melanoma, immune checkpoint blockade with monoclonal antibodies directed at the inhibitory immune receptors CTLA-4 and PD-1 demonstrated considerable synergistic activity [20]. The objective-response rate for all patients receiving varying intravenous doses of nivolumab and ipilimumab every 3 weeks was 40%. However, grade 3 or 4 adverse events related to therapy occurred in 53% of patients in the concurrent-regimen group.

Similarly, the combination of ipilimumab and nivolumab is currently being assessed in phase I trials in patients with various solid tumors, including SCLC, renal cell carcinoma and colon cancer. Interim data from another phase I study evaluating first-line nivolumab plus ipilimumab in 46 advanced NSCLC patients has recently been presented [21]. Similarly to the melanoma trial, varying doses of the two checkpoint inhibitors were used. The overall response rate was 22% while the median duration of response was not reached. Again, the toxicity was higher than seen in monotherapies with either drug alone and included 3 treatment-related deaths due to respiratory failure, bronchopulmonary hemorrhage and toxic epidermal necrolysis. Therefore, the most appropriate dose and schedule for the nivolumab and ipilimumab combination is still to be defined. Further trials are ongoing evaluating dual CTLA-4 and PD-1 checkpoint pathway inhibition, including tremelimumab or ipilimumab (anti-CTLA-4), pembrolizumab (anti-PD-1), and MEDI-4736 (anti-PD-L1).

Several other potential immunotherapeutic combinations are being investigated in solid tumors, including in patients with NSCLC. One example is the anti-LAG-3 monoclonal antibody (BMS-986016), which is currently being evaluated in combination with nivolumab in a phase I clinical trial in patients with solid tumors. Other clinical trials are evaluating combination regimens that incorporate lirilumab (anti-KIR antibody), which may potentiate innate immunity, with either nivolumab or ipilimumab in patients with advanced solid tumors.

Finally, combinations of cancer vaccines to generate antitumor T cells and immune checkpoint inhibitors to prevent T cell anergy have been explored. Cancer vaccines consist of various strategies that intend to prime and boost specific immune responses in a tumor-bearing host. This hypothesis was tested in a small trial on 11 patients with metastatic melanoma receiving ipilimumab after vaccination with autologous granulocyte-macrophage colony-stimulating factor (GM-CSF)-secreting

Table 1. Ongoing clinical trials with immune checkpoint inhibitors in lung cancer

Drug	Target	Phase	Population	Design	Primary outcome measure	NCT number	Completion date
Ipilimumab	CTLA-4	1	NSCLC	Ipilimumab + neoadjuvant chemotherapy vs. neoadjuvant chemotherapy	% of subjects with detectable tumor-specific circulating T cells	NCT01820754	May 2018
		1	AML NSCLC various	Ipilimumab in patients previously vaccinated with GM-CSF-based autologous tumor vaccines	Toxicity	NCT00039091	November 2007
		1	NSCLC	Ipilimumab + erlotinib or crizotinib	Safety	NCT01998126	December 2015
		2	SCLC	Ipilimumab + carboplatin/etoposide	PFS	NCT01331525	May 2015
		3	Metastatic/ED SCLC	Ipilimumab plus etoposide/platinum vs. etoposide/platinum	OS	NCT01450761	March 2017
		3	NSCLC SCC	Ipilimumab + paclitaxel/carboplatin vs. paclitaxel/carboplatin	OS	NCT01285609	December 2016
Tremelimumab	CTLA-4	1	Solid tumors	MEDI-4736 + tremelimumab	AE rate	NCT01975831	December 2016
		2	NSCLC	Gefitinib, AZD9291, tremelimumab or selumetinib + docetaxel with a equential switch to MEDI4736	OR	NCT02179671	May 2016
Nivolumab	PD-1	1	Solid tumors	Nivolumab + interleukin-21	AE rate/severity	NCT01629758	April 2015
		1	NSCLC	Safety of nivolumab + various standard chemotherapies	AE/worsttoxicity grade/lab test abnormalities	NCT01454102	August 2015

Table 1. Continued

Drug	Target	Phase	Population	Design	Primary outcome measure	NCT number	Completion date
		1/2	Advanced/meta-static solid tumors including NSCLC	Nivolumab or nivolumab + ipilimumab	ORR	NCT01928394	April 2015
		2	Advanced/meta-static SCC NSCLC who have received >2 previous treatments	Nivolumab as third-line onwards therapy	ORR	NCT01721759	February 2015
		3	SCC NSCLC after failure of platinum	Nivolumab vs. docetaxel as second-line therapy	ORR and OS	NCT01642004	August 2015
		3	NSCC NSCLC after failure of platinum	Nivolumab vs. docetaxel as second-line therapy	ORR and OS	NCT01673867	November 2015
		3	NSCLC	Nivolumab vs. investigator's choice of chemotherapy	PFS	NCT02041533	January 2018
		3	NSCLC	Nivolumab second-line onwards	Safety	NCT02066636	March 2019
Pembrolizumab (MK3475)	PD-1	1	Advanced solid tumors including NSCLC	MK3475 + cisplatin/pemetrexed or carboplatin/paclitaxel	DLT rate	NCT01840579	April 2015
		1/2	NSCLC	Pembrolizumab + paclitaxel/carboplatin/bevacizumab/pemetrexed/ipilimumab/erlotinib/gefitinib	ORR, PFS	NCT02039674	June 2019
		2/3	NSCLC progressed after platinum	Low or high dose of MK3475 vs. docetaxel	OS, PFS, AE rate	NCT01905657	January 2020

Table 1. Continued

Drug	Target	Phase	Population	Design	Primary outcome measure	NCT number	Completion date
		3	NSCLC	Pembrolizumab vs. paclitaxel/carboplatin/pemetrexed/cisplatin/gemcitabine	PFS	NCT02142738	December 2017
MPDL3280A	PD-L1	1	Advanced solid tumors	MPDL3280A + bevacizumab + various other chemotherapies	AE/DLT rate	NCT01633970	July 2015
		1	NSCLC	MPDL3280A + erlotinib	Safety	NCT02013219	August 2016
		2	Advanced/meta-static NSCLC after platinum failed	MPDL3280A vs. docetaxel as second-line	OS	NCT01903993	March 2017
		2	Advanced/meta-static PD-1+ NSCLC	MPDL3280A	OR	NCT01846416	May 2015
		2	Advanced/meta-static PD-1+ NSCLC	MPDL3280A	OR	NCT02031458	March 2018
		3	NSCLC	MPDL3280A vs. docetaxel	OS	NCT02008227	June 2018
MEDI-4736	PD-L1	1	NSCLC	MEDI-4736 + tremelimumab	Safety	NCT02000947	January 2017
		1	NSCLC	MEDI-4736 + gefitinib	Safety	NCT02088112	October 2017
		2	NSCLC	MEDI-4736 third-line	OR	NCT02087423	August 2016
		3	NSCLC	MEDI-4736 following concurrent chemoradiation	OR, PFS	NCT02125461	November 2020

Table 1. Continued

Drug	Target	Phase	Population	Design	Primary outcome measure	NCT number	Completion date
Lirilumab	KIR	1	Advanced/meta-static solid tumors	Lirilumab + nivolumab	AE rate/lab test abnormalities	NCT01714739	September 2015
		1	Advanced/meta-static solid tumors	Lirilumab + ipilimumab	AE rate	NCT01750580	July 2015
BMS-986016	LAG-3	1	Advanced solid tumors	BMS-986016 alone or + nivolumab	AE rate/lab test abnormalities	NCT01968109	October 2020
Bavituximab	PS	1	Previously untreated stage IV NSCLC	Bavituximab + pemetrexed/ carboplatin in first-line therapy	Severe side effects	NCT01323062	May 2018
		3	Non-squamous NSCLC	Bavituximab + docetaxel vs. docetaxel	OS	NCT01999673	December 2016

ED = Extensive disease; AML = acute myeloid leukemia; SCC = squamous cell carcinoma; NSCC = non-squamous cell carcinoma; OS = overall survival; AE = adverse effect; ORR = objective response rate; DLT = dose limiting toxicity. Adapted from Reck and Paz-Arez [18].

tumor cells [22]. Interestingly, 8 of the 11 subjects displayed meaningful antitumor effects, including three partial responses. The extent of therapy-induced tumor necrosis was linearly related to the natural logarithm of the ratio of intratumoral CD8+ effector T cells to FoxP3+ regulatory T cells in posttreatment biopsies. All patients demonstrated mild side effects that were related to the inflammatory response without grade 3 or 4 toxicities. Similarly, in a preclinical study, vaccination with irradiated B16 melanoma cells expressing either GM-CSF or the Flt3 ligand combined with CTLA-4 antibody blockade promoted rejection of preimplanted melanoma tumors. By the combination of CTLA-4 and PD-1 blockade an even more effective tumor rejection and increased effector T cell infiltration was noted [23]. Moreover, the combination of PD-1 blockade with GM-CSF-secreting tumor cell immunotherapy led to increased overall survival and decreased tumor growth in melanoma and colon carcinoma tumor models compared with either monotherapy [24]. These data suggest that combining agonistic antibodies delivering a costimulatory signal with antibodies blocking inhibitory signals may promote potent antitumor immunity. Currently, a phase I trial investigates anti-CTLA-4 therapy in patients previously vaccinated with GM-CSF-based autologous tumor vaccines and harboring various malignancies, including NSCLC. In addition, several other vaccines are currently undergoing clinical trials in NSCLC (including belagenpumatucel-L, GV1001, TG4010 and GSK1572932A), which may be candidate therapeutic partners with a checkpoint inhibitor.

Combination of Immunotherapy with Targeted Therapy

Targeted therapies act by blocking essential signaling pathways or receptors leading to tumor regression, particularly in those tumors that are dependent on the proper functioning of the respective target. Hence, molecular characterization of all tumors is warranted to preselect those patients that are most likely to benefit from targeted therapies. Still, clinical success is somewhat limited and tumor regrowth is finally seen in most patients. To date, a host of so-called targeted agents have been developed and are being explored in clinical trials.

Due to the multiplicity of targets (e.g. a specific growth receptor or a defined signaling pathway), multiple rationales for interactions with the immunosurveillance have been defined [25]. So far, largely preclinical data suggest that the interplay of targeted agents and immunotherapy is complex and may be affected by various mechanisms. Similarly to cytotoxic therapy, tumor regression may consequently lead to a decrease in tumor-associated immunosuppression by abrogating the production of tumorigenic inflammation and by inhibiting immunosuppressive cell types supporting a combinational approach with immunostimulating agents. By decreasing the amount of tumor cells, tumor cell senescence and clearance by T cells may be facilitated. Moreover, the release of large amounts of antigenic debris on tumor cell death may further stimulate immunogenic recognition by antigen-representing cells such as dendritic cells [25]. In addition to tumor cell killing, targeted therapies may also directly modulate immune responses. Several receptor-targeting agents also affect receptor tyrosine kinases that are expressed on MDSCs (myeloid-derived suppressor cells), such as c-KIT and VEGFR-1 (vascular endothelial growth factor receptor 1), making these promising immunomodulators [26].

In a murine colon carcinoma model, sunitinib given before a vaccine increased intratumoral infiltration of antigen-specific T lymphocytes, decreased regulatory T cells and MDSCs, reduced tumor volume and increased survival [27]. In another study, established melanoma cell lines as well as primary melanoma cells from tumor biopsies, which were notoriously resistant toward T cell lysis, became sensitive upon bortezomib treatment, indicating that some targeted therapies might sensitize tumor cells to immune-mediated destruction [28]. Also, treatment with the mTOR inhibitor ra pamycin effectively sensitized established cancers to adoptive immunotherapy in vivo [29]. Finally, monoclonal antibodies may facilitate the uptake of tumor antigens by dendritic cells. Treatment of human colon cancer cell lines with cetuximab, a human/mouse chimeric antibody able to bind the epidermal growth factor receptor (EGFR), promoted the phagocytosis of untreated or chemotherapy-treated colon cancer cells by human dendritic cells, which was associated with dendritic cell maturation and enhanced the induction of tumor-specific T cells [30].

In contrast to these positive interactions, some targeted therapies with immunostimulatory potential may also display immunosuppressive activities. For example, hsp90 (heat shock protein 90) proteins are considered to be key components of immune function since a number of key signaling proteins that are hsp90 clients also play important roles in the activation of normal cells involved in innate and adaptive immune responses. Inhibition of hsp90 has been correlated with enhanced antigen presentation on MHC class I molecules and augmented tumor cell recognition by CD8+ T cells in vitro [31]. In contrast, in other studies, blocking hsp90 function resulted in a decreased CD4+ T cell activation and proliferation, leading to suppression of an inflammatory response [32].

In lung cancer, only limited clinical data has been reported so far. Recently, interim results from a phase I study evaluating nivolumab and erlotinib in a cohort of 21 NSCLC patients harboring activating EGFR mutations have been presented. While the majority of patients were pretreated with EGFR TKIs (only 1 patient was EGFR TKI naïve), a response rate of 19% and progression-free rate at 24 weeks of 47% could be reached [33].

To date, targeted therapies offer great promise for boosting responses with immunotherapy. Still, various questions regarding the appropriate timing, dosage and the sequence of administration of targeting agents and immunomodulators remain to be addressed. Moreover, the toxicity profiles are still to be determined. Finally, the selection of the appropriate target and agent (e.g. monoclonal antibody or small molecule) for combinations with immunomodulating therapy will probably be crucial to the success of combinatorial approaches. A more complete understanding of the effect of targeted therapies on specific molecular pathways in immune cells should help to advance the integration of these agents with immunotherapy.

Combination of Immunotherapy with Antiangiogenic Therapy

When combining targeted agents with immunotherapy, antiangiogenic agents are of particular interest. Angiogenesis is the growth of new microvessels from the preexisting vasculature, a process that involves a fine balance of proangiogenic and antiangiogenic factors and a fine orchestrated coordination between multiple cell types such as macrophages, endothelial cells and pericytes [34]. Supported by an abnormal tumor microenvironment with interstitial hypertension, hypoxia and acidosis, both tumor and stromal cells contribute to a shift of this balance towards tumor angiogenesis, leading to a new vasculature formation that is necessary for cancer cells to proliferate beyond microscopic size and to metastasize. The vascular endothelial growth factor (VEGF) plays a key role in regulation, both in normal and cancer cells, promoting the endothelial cell migration and proliferation necessary for angiogenesis.

In addition, VEGF also manifests potent immunoregulatory actions, such as inhibition of the differentiation and maturation of dendritic cells from hematopoietic progenitors. In mouse tumor models, treatment of established tumors with an anti-VEGF antibody did not affect the rate of tumor growth but significantly improved the number and function of host dendritic cells [35]. A combined treatment with anti-VEGF antibodies and activated dendritic cells produced a prolonged and much more pronounced antitumor effect [35]. In a different study, VEGF impaired the differentiation of human monocytes to dendritic cells which was reversed by the treatment of bevacizumab and sorafenib but not sunitinib [36]. By contrast, bevacizumab had limited effects when added to matured dendritic cells, suggesting that there is a specific window during dendritic cell development in which VEGF exerts regulatory functions. Moreover, neither bevacizumab nor sorafenib were able to reverse the inhibitory effects of supernatants of hypoxic renal cancer cells on dendritic cell maturation, suggesting that VEGF is not the only immunomodulatory factor secreted by tumor cells [36]. Still, the administration of a VEGF antibody combined with adoptive T cell transfer enhanced the growth inhibition of established melanoma tumors, infiltration of T cells and prolonged survival compared with either monotherapy [37].

Moreover, endothelial cells also contribute to immune regulation by both limiting and facilitating the trafficking of various immune cell populations out of the vasculature and into tissue [3]. Depending on the microenvironment they can also be induced to stimulate immune reactivity or to be immune inhibitory. For example, in response to pathogens such as bacterial lipoproteins, endothelial cells become activated and produce cytokines such as IL-6, IL-8 and G-CSF with subsequent neutrophil recruitment [38]. In contrast, endothelial cells may also act as immune suppressive cells through both cell contact-dependent and cell contact-independent mechanisms. For example, a physiologically beneficial effect of immune inhibitory endothelial cells is to maintain the blood-brain barrier by limiting T cell transmigration into the central nervous system [3]. However, the immunomodulatory role of endothelial cells in tumors has not been well studied. In one study, expression of Egfl7, a secreted protein specifically expressed by endothelial cells in normal organs, was also expressed by lung carcinoma cells and promoted tumor growth and metastasis in immunocompetent mice but not in immunodeficient mice [39]. Moreover, Egfl7 inhibited the expression of leukocyte adhesion molecules by endothelial cells, thereby preventing lymphocyte adhesion.

In a phase I study, 46 patients with metastatic melanoma were treated in four dosing cohorts with the combination of ipilimumab and the anti-VEGF antibody bevacizumab. With an overall response and disease control rate of 19.6 and 67.4%, respectively, the median survival was 25.1 months [40]. Moreover, bevacizumab influenced changes in tumor vasculature and immune responses with ipilimumab administration. With 13 patients having reported treatment-related grade 3 or 4 adverse events (28.3%), inflammatory toxicities were generally higher than expected with ipilimumab alone, but remained manageable. In NSCLC, current trials are investigating various combinations of immunomodulatory therapies and systemic therapy, including bevacizumab (NCT02039674) as well as nivolumab and bevacizumab maintenance therapy (NCT01454102).

Side Effects

One key issue for combinatorial therapies is whether intensified antitumor effects can be achieved without a corresponding increase in serious toxicities. Clinical studies investigating immunotherapeutic agents as monotherapy have demonstrated an acceptable safety profile in general, particularly compared with conventional cytotoxics. Most immune-related adverse events seemed to be reversible and inflammatory in nature, and included diarrhea/colitis and pneumonitis. However, safety and efficacy need to be further characterized in larger trials.

Combining antitumor agents usually results in combining and adding toxicities known from the respective drugs as monotherapies. For example, a phase II trial evaluating ipilimumab combined with platinum-based chemotherapy in NSCLC demonstrated similar nonhematologic adverse events typically associated with carboplatin and paclitaxel [14]. Interestingly, for other adverse events commonly associated with immunotherapy, such as rash, pruritus and diarrhea, there was a trend toward increased incidence in the ipilimumab groups. The overall incidence of grade 3 and 4 immune-related adverse events was 15, 20 and 6% for the phased ipilimumab, concurrent ipilimumab and control groups, respectively [14].

Similarly, a trial of 56 NSCLC patients treated with platinum-based doublet chemotherapy and nivolumab experienced grade 3–4 treatment-related adverse events in 45% of all patients, including pneumonitis (7%), and fatigue and acute renal failure (5% each) [17]. No dose-limiting toxicities were seen during the first 6 weeks of treatment.

When adding nivolumab to erlotinib in 21 EGFR-mutated patients, interim results from a phase I study displayed grade 3–4 treatment-related adverse events in 4 patients, including elevated liver enzymes and diarrhea [33]. No pneumonitis of any grade was observed.

By contrast, combinations of multiple immunomodulating agents or with anti-angiogenic drugs may be associated with increased toxicity. For example, a phase I study demonstrated some considerable inflammatory toxicity in metastatic melanoma patients treated with ipilimumab plus bevacizumab, including 28% of patients having reported treatment-related grade 3 or 4 adverse events [40]. Recently, interim results have been published from a multicenter phase Ib study evaluating the combination of an anti-PDL1 antibody (MPDL3280A) with bevacizumab with or without oxaliplatin-containing chemotherapy in patients with various locally advanced or metastatic solid malignancies [41]. While both combinations were generally well tolerated and no unexpected toxicity was observed, grade 3/4 adverse events regardless of attribution were noted in 49 and 64% of patients, respectively, and included abdominal pain, hyperbilirubinemia, pneumonia, neutropenia and diarrhea. Moreover, the combination of ipilimumab and nivolumab has shown some toxicity in melanoma and lung cancer patients that exceeded that of either single agent [20, 21]. With about 50% of all patients reporting with grade 3/4 treatment-related adverse events, side effects of interest included pneumonitis (6%), diarrhea (8%) and colitis (8%) [21]. However, most toxicities were manageable in these trials, and the number of patients was very limited.

Conclusion and Outlook

Limited data from current studies have suggested that combinations with immunomodulating agents are feasible and may result in enhanced antitumor efficacy. Furthermore, toxicities may be somewhat enhanced but seem to be manageable. The interplay of targeted agents and immunotherapy is complex, however, and parameters such as timing, dosage and sequence of administration are all likely to influence the overall antitumor effects and toxicity profiles of the combined treatments. A more complete understanding of the effect of targeted therapies on specific molecular pathways in immune cells should help to advance the integration of these agents with immunotherapy. Multiple clinical studies are currently recruiting so that a more concise understanding about the clinical benefit of these approaches will be more possible in the near future.

References

1 Hurwitz AA, Watkins SK: Immune suppression in the tumor microenvironment: a role for dendritic cell-mediated tolerization of T cells. Cancer Immunol Immunother 2012;61:289–293.

2 Anderson MJ, Shafer-Weaver K, Greenberg NM, et al: Tolerization of tumor-specific T cells despite efficient initial priming in a primary murine model of prostate cancer. J Immunol 2007;178:1268–1276.

3 Young MR: Endothelial cells in the eyes of an immunologist. Cancer Immunol Immunother 2012;61:1609–1616.

4 Pardoll DM: The blockade of immune checkpoints in cancer immunotherapy. Nat Rev Cancer 2012;12:252–264.

5 Arlen PM, Gulley JL, Parker C, et al: A randomized phase II study of concurrent docetaxel plus vaccine versus vaccine alone in metastatic androgen-independent prostate cancer. Clin Cancer Res 2006;12:1260–1269.

6 Zitvogel L, Apetoh L, Ghiringhelli F, et al: Immunological aspects of cancer chemotherapy. Nat Rev Immunol 2008;8:59–73.

7 Apetoh L, Ghiringhelli F, Tesniere A, et al: Toll-like receptor 4-dependent contribution of the immune system to anticancer chemotherapy and radiotherapy. Nat Med 2007;13:1050–1059.

8 Zielinski C, Knapp S, Mascaux C, et al: Rationale for targeting the immune system through checkpoint molecule blockade in the treatment of non-small-cell lung cancer. Ann Oncol 2013;24:1170–1179.

9 Ramakrishnan R, Assudani D, Nagaraj S, et al: Chemotherapy enhances tumor cell susceptibility to CTL-mediated killing during cancer immunotherapy in mice. J Clin Invest 2010;120:1111–1124.

10 Peggs KS, Quezada SA, Chambers CA, et al: Blockade of CTLA-4 on both effector and regulatory T cell compartments contributes to the antitumor activity of anti-CTLA-4 antibodies. J Exp Med 2009;206:1717–1725.

11 Waterhouse P, Penninger JM, Timms E, et al: Lymphoproliferative disorders with early lethality in mice deficient in CTLA-4. Science 1995;270:985–988.

12 Klein O, Ebert LM, Nicholaou T, et al: Melan-A-specific cytotoxic T cells are associated with tumor regression and autoimmunity following treatment with anti-CTLA-4. Clin Cancer Res 2009;15:2507–2513.

13 Ribas A, Camacho LH, Lopez-Berestein G, et al: Antitumor activity in melanoma and anti-self responses in a phase I trial with the anti-cytotoxic T lymphocyte-associated antigen 4 monoclonal antibody CP-675,206. J Clin Oncol 2005;23:8968–8977.

14 Lynch TJ, Bondarenko I, Luft A, et al: Ipilimumab in combination with paclitaxel and carboplatin as first-line treatment in stage IIIB/IV non-small-cell lung cancer: results from a randomized, double-blind, multicenter phase II study. J Clin Oncol 2012;30:2046–2054.

15 Reck M, Bondarenko I, Luft A, et al: Ipilimumab in combination with paclitaxel and carboplatin as first-line therapy in extensive-disease-small-cell lung cancer: results from a randomized, double-blind, multicenter phase 2 trial. Ann Oncol 2012;24:75–83.

16 Zatloukal P, Heo DS, Park K, et al: Randomized phase II clinical trial comparing tremelimumab (CP-675,206) with best supportive care (BSC) following first-line platinum-based therapy in patients (pts) with advanced non small cell lung cancer (NSCLC) [abstract]. J Clin Oncol 2009;27:8071.

17 Antonia SA, Brahmer JR, Gettinger SN, et al: Nivolumab (anti-PD-1; BMS-936558, ONO-4538) in combination with platinum-based doublet chemotherapy (PT-DC) in advanced non-small cell lung cancer (NSCLC) [abstract]. J Clin Oncol 2014;32:8113.

18 Reck M, Paz-Ares L: Immunologic checkpoint blockade in lung cancer. Semin Oncol 2015;42:402–417.

19 Terme M, Ullrich E, Aymeric L, et al: IL-18 induces PD-1-dependent immunosuppression in cancer. Cancer Res 2011;71:5393–5399.

20 Wolchok JD, Kluger H, Callahan MK, et al: Nivolumab plus ipilimumab in advanced melanoma. N Engl J Med 2013;369:122–133.

21 Antonia SA, Gettinger SN, Chow LQM, et al: Nivolumab (anti-PD-1; BMS-936558, ONO-4538) and ipilimumab in first-line NSCLC: interim phase I results (abstract). J Clin Oncol 2014;32:8023.

22 Hodi FS, Butler M, Oble DA, et al: Immunologic and clinical effects of antibody blockade of cytotoxic T lymphocyte-associated antigen 4 in previously vaccinated cancer patients. Proc Natl Acad Sci U S A 2008;105:3005–3010.

23 Curran MA, Montalvo W, Yagita H, et al: PD-1 and CTLA-4 combination blockade expands infiltrating T cells and reduces regulatory T and myeloid cells within B16 melanoma tumors. Proc Natl Acad Sci U S A 2010;107:4275–4280.

24 Li B, VanRoey M, Wang C, et al: Anti-programmed death-1 synergizes with granulocyte macrophage colony-stimulating factor – secreting tumor cell immunotherapy providing therapeutic benefit to mice with established tumors. Clin Cancer Res 2009;15:1623–1634.

25 Vanneman M, Dranoff G: Combining immunotherapy and targeted therapies in cancer treatment. Nat Rev Cancer 2012;12:237–251.

26 Kao J, Ko EC, Eisenstein S, et al: Targeting immune suppressing myeloid-derived suppressor cells in oncology. Crit Rev Oncol Hematol 2010;77:12–19.
27 Farsaci B, Higgins JP, Hodge JW: Consequence of dose scheduling of sunitinib on host immune response elements and vaccine combination therapy. Int J Cancer 2011;130:1948–1959.
28 Seeger JM, Schmidt P, Brinkmann K, et al: The proteasome inhibitor bortezomib sensitizes melanoma cells toward adoptive CTL attack. Cancer Res 2010; 70:1825–1834.
29 Hahnel PS, Thaler S, Antunes E, et al: Targeting AKT signaling sensitizes cancer to cellular immunotherapy. Cancer Res 2008;68:3899–3906.
30 Correale P, Botta C, Cusi MG, et al: Cetuximab ± chemotherapy enhances dendritic cell-mediated phagocytosis of colon cancer cells and ignites a highly efficient colon cancer antigen-specific cytotoxic T-cell response in vitro. Int J Cancer 2011;130:1577–1589.
31 Kawabe M, Mandic M, Taylor JL, et al: Heat shock protein 90 inhibitor 17-dimethylaminoethylamino-17-demethoxygeldanamycin enhances EphA2$^+$ tumor cell recognition by specific CD8$^+$ T cells. Cancer Res 2009;69:6995–7003.
32 Yun TJ, Harning EK, Giza K, et al: EC144, a synthetic inhibitor of heat shock protein 90, blocks innate and adaptive immune responses in models of inflammation and autoimmunity. J Immunol 2010; 186:563–575.
33 Rizvi NA, Chow LQM, Borghaei H, et al: Safety and response with nivolumab (anti-PD-1; BMS-936558, ONO-4538) plus erlotinib in patients (pts) with epidermal growth factor receptor mutant (EGFR MT) advanced NSCLC (abstract). J Clin Oncol 2014;32: 8022.
34 Carmeliet P, Jain RK: Molecular mechanisms and clinical applications of angiogenesis. Nature 2011; 473:298–307.
35 Gabrilovich DI, Ishida T, Nadaf S, et al: Antibodies to vascular endothelial growth factor enhance the efficacy of cancer immunotherapy by improving endogenous dendritic cell function. Clin Cancer Res 1999;5:2963–2970.
36 Alfaro C, Suarez N, Gonzalez A, et al: Influence of bevacizumab, sunitinib and sorafenib as single agents or in combination on the inhibitory effects of VEGF on human dendritic cell differentiation from monocytes. Br J Cancer 2009;100:1111–1119.
37 Shrimali RK, Yu Z, Theoret MR, et al: Antiangiogenic agents can increase lymphocyte infiltration into tumor and enhance the effectiveness of adoptive immunotherapy of cancer. Cancer Res 2010;70: 6171–6180.
38 Wilhelmsen K, Mesa KR, Prakash A, et al: Activation of endothelial TLR2 by bacterial lipoprotein upregulates proteins specific for the neutrophil response. Innate Immun 2011;18:602–616.
39 Delfortrie S, Pinte S, Mattot V, et al: Egfl7 promotes tumor escape from immunity by repressing endothelial cell activation. Cancer Res 2011;71:7176–7186.
40 Hodi FS, Lawrence D, Lezcano C, et al: Bevacizumab plus ipilimumab in patients with metastatic melanoma. Cancer Immunol Res 2014;2:632–642.
41 Lieu C, Bendell J, Powderly J, et al: Safety and efficacy of MPDL3280A (anti-PDL1) in combination with bevacizumab (bev) and/or chemotherapy (chemo) in patients (pts) with locally advanced or metastatic solid tumors (abstract). Ann Oncol 2014;25: 10490.

Priv.-Doz. Dr. med. Martin Reck
Department of Thoracic Oncology
LungenClinic Großhansdorf, Wöhrendamm 80
22927 Großhansdorf (Germany)
E-Mail m.reck@lungenclinic.de

Michielin O, Coukos G (eds): Immuno-Oncology.
Prog Tumor Res. Basel, Karger, 2015, vol 42, pp 95–109 (DOI: 10.1159/000437179)

Promise of Immunotherapy in Lung Cancer

Anil Shanker[a, b] • Mikhail M. Dikov[c] • David P. Carbone[c]

[a]Department of Biochemistry and Cancer Biology, Meharry Medical College School of Medicine, and [b]Host-Tumor Interactions Research Program, Vanderbilt-Ingram Cancer Center, Vanderbilt University, Nashville, Tenn., and [c]Division of Medical Oncology, Department of Internal Medicine, Ohio State University Medical Center, Columbus, Ohio, USA

Abstract

Metastatic lung cancer is the most common cause of cancer mortality globally in both men and women, with 5-year survival of less than 5%. Standard treatment approaches for metastatic lung cancer are based on chemotherapy, with radiation and surgery used for local control, but these rarely result in relapse-free survivals longer than 2–3 years, although they may provide symptom relief. Thus, additional tools are needed to treat this disease. In this chapter, we discuss the various immune-based cancer treatments for lung cancer patients that are being developed, and the increasing awareness that therapies targeted at overcoming immune evasion mechanisms may be essential to clinical efficacy.

Approximately one in three people in the world is afflicted with cancer in his or her lifetime. However, the majority of people, including those who are heavy smokers, still remain cancer-free throughout a long life-span. Heavy smokers bombard their lungs with carcinogens daily, accumulating somatic mutations that result in novel protein sequences and tumour-promoting mutations in airway epithelial cells, yet most of these individuals remain cancer-free due to a variety of mechanisms that resist the progression of these preneoplastic lesions into clinically evident cancer. These mechanisms range from simple contact inhibition by neighbouring stromal cells to elaborate molecular mechanisms of DNA repair, suppression of oncogene activation, tumour suppressor genes, epigenetic stabilisation of chromatin, apoptosis and immune surveillance.

Recent cancer therapy studies have built on observations that immune cells can shape the outcome of various anticancer approaches. As such, immune cells and their molecular mediators have evolved into bona fide players in cancer therapy. Indeed, besides melanoma and renal cancer, which respond to the systemic administration of immune-targeted biologics and show signals of response to certain therapeutic vaccines, lung cancer, which heretofore had given clinicians little indication of immune sensitivity, is another major cancer type for which new immune-based therapies are showing significant promise. In spite of this, the history of cancer immunotherapy attempts is long, and dates back to the late 1800s with the development of bacterial extracts to stimulate the immune system by William Coley, and in more modern times was reinvigorated in 1975 when Picibanil, a lyophilized preparation of *Streptococcus pyogenes* that acts as a TLR2/TLR4 agonist, was licensed for use in lung cancer patients by the Japanese Ministry of Health and Welfare [1].

Lung cancer is the most common cause of cancer mortality globally, representing 13% of all cancer diagnoses each year and nearly 1 in 5 cancer-related deaths. The majority of lung cancer patients are diagnosed with advanced disease (stage IIIb/IV). The two major forms of lung cancer are non-small cell lung cancer (NSCLC) and small cell lung cancer (SCLC). NSCLC comprises approximately 85% of all lung cancers. Of these, approximately 25–30% are squamous NSCLC, while the rest are non-squamous. Survival rates for metastatic lung cancer including NSCLC and SCLC are poor, with 5-year survival of less than 5% [2]. Available treatment approaches, including surgery, chemotherapy and radiotherapy, are unlikely to result in a relapse-free cure, although they may improve survival and provide symptom relief in lung cancer patients. In this chapter, we review various strategies that lung cancer cells use to evade immune response and highlight the new immune-based cancer treatments for lung cancer patients that are being developed.

Immune Evasion of Lung Cancer

It is clear that clinically evident tumours have evaded an effective immune response, and basic immunology studies are defining a delicate interplay between natural immunoactivation and immunosuppression mechanisms that are overcome or usurped in patients who develop cancers. As can be seen in patients with raging autoimmunity and graft-versus-host disease, the immune system is very powerful, and as such has evolved robust and very complex regulatory mechanisms to keep it in check. These mechanisms include the local or systemic secretion of soluble factors, generation of specific regulatory cell populations, and direct cell-cell interactions between immune cells and each other, the host and the tumour.

Suppressive Cellular Mechanisms

In NSCLC as well as in SCLC and mesothelioma, the role of increased immunosuppressive cells downregulating immune activation is high. This involves generation of a highly immunosuppressive tumour microenvironment by tumour-derived factors and differentiation of T regulatory (Treg) cells and myeloid-derived suppressor cells. In addition, high concentrations of adenosine in tumour microenvironments inhibit antitumour cytotoxic lymphocyte responses. As one example, although T cells express inhibitory adenosine A2A receptors (A2AR) that suppress their activation and inhibit immune killing of tumours, myeloid cell A2ARs primarily suppress the immune response to tumours. The growth of transplanted syngeneic Lewis lung carcinoma cells is slowed in Adora2a$^{f/f}$-LysMCre$^{+/-}$ mice, which selectively lack myeloid A2ARs [3]. Myeloid-selective A2AR deletion is associated with increased numbers and activation of natural killer (NK) cells and antigen-specific CD8+ T cells in lung infiltrates.

The results of one study suggest that serum of NSCLC patients may exert an inhibitory effect on the cytotoxicity of immune effector cells such as NK cells, and this suppression may be further enhanced with chemotherapy [4]. Moreover, the expression of NK cell inhibitory receptors such as NKG2A increase remarkably in a lung cancer microenvironment, with downregulation of the expression of the NK cell-activating receptors NKG2D and Ly49I [5]. In another study, the frequency and functional activity of Treg cells was analysed in 31 NSCLC stage IB-IIIA patients treated by neoadjuvant cetuximab/docetaxel/cisplatin prior to surgery. Results show that Treg levels at diagnosis are not linked to an inferior clinical response to chemoimmunotherapy in NSCLC in vivo, even though Treg cells potently inhibit NK-mediated and cetuximab-induced antibody-induced cellular cytotoxicity in vitro [6]. Lung cancer is also associated with decreased expression of granzyme B, perforin and IFN-γ by tumour-infiltrating T cells, NK T-like and NK cells, possibly as a result of soluble factors produced by the cancer cells, including prostaglandin E2 (PGE2) via the PGE2/COX2 pathway [7]. Tumour-derived transforming growth factor-β1 (TGF-β) is also broadly immunosuppressive. TGF-β-treated human NK cells exhibit reduced tumour lysis and abrogated perforin polarization to the immune synapse by loss of surface expression of activating killer Ig-like receptor 2DS4 and NKp44, despite the intact cytoplasmic levels of these receptors [8]. Moreover, by inducing microRNA-183, TGF-β can repress transcription/translation of DNAX-activating protein 12 kDa (DAP12), which is critical for surface NK receptor stabilization and downstream signal transduction. Reduced DAP12 expression in tumour-associated NK cells is observed in lung cancer patients [8].

Soluble Factors

A highly immunosuppressive tumour microenvironment is generated as a result of soluble factors produced by the cancer cells. Since the first demonstration that TGF-β1 secreted by immunogenic tumours may promote tumour escape from T cell cytolytic immune response [9] by its pleiotropic effects on T cell proliferation, differentiation, cytokine production and antigen presentation, an impressive variety of immunosuppressive and tumour-promoting cytokines or factors, ranging from acute phase reactants with non-specific inhibitory properties to adhesion molecules blocking cell interactions or apoptosis, have been reported. Tumours may escape from immune detection by producing GM-CSF and soluble vascular endothelial growth factor (VEGF), thereby disrupting the balance of cytokines needed for the maturation of fully functional antigen-presenting cells [10, 11]. In patients with metastatic NSCLC, overproduction of VEGF showed a statistically positive correlation with IL-4, suggesting that

VEGF might be responsible for the predominance of serum Th2 cytokines, limiting the efficacy of dendritic cell vaccination and cytokine-induced killer (CIK) cell therapy [12]. We also identified that the elevated circulating VEGF was one of the causal factors for tumour-associated defects in T lymphocytes mediated by alteration of the expression pattern of Notch ligands and reduced Notch signalling in the haematopoietic microenvironment [13–15]. Lung tumour cells can also promote their growth by exploiting the low-affinity A2B adenosine receptor-dependent upregulation of VEGF in the tumour microenvironment [16, 17]. In addition, generation of adenosine by granulocytic myeloid-derived suppressor cells can promote their expansion and facilitate their immunosuppressive activity in the lung cancer tumour microenvironment [18].

IL-6 and IL-10 are other major immune dysregulating cytokines elevated in various haematological malignancies [19, 20] and carcinomas [21]. Tumours as a result of their hypoxic metabolism, for example by elevated adenosine, can inhibit IL-12 and stimulate IL-10 production by monocytes [22, 23]. Dendritic cells exposed to IL-10 induce anergy in tumour-specific T cells and prime IL-4-secreting T cells [24]. Another cytokine, IL-13, produced by NKT cells, signals through the IL-4 receptor-STAT6 pathway, and downregulates T cell antitumour responses, facilitating tumour recurrence [25]. Another feature of the tumour microenvironment is elevated levels of chemokine CCL22, which attracts CCR4-expressing Treg cells capable of suppressing an effective antitumour immune response [26, 27]. A large majority of human tumours also constitutively express indoleamine 2,3-dioxygenase, which catabolizes the essential amino acid tryptophan [28]. Tryptophan deprivation sensitizes activated T cells to apoptosis [29], thereby eliminating antitumour T cells [30].

Recent advances in the understanding of mechanisms of the immune system in tumour immunosurveillance have resulted in the recognition that tumours can evade their destruction via the dysregulation of immunostimulatory, coinhibitory or checkpoint signals, which are described in detail below. This has led to the development of approved immunotherapeutic approaches for lung cancer therapy for the first time.

Attempts to use the immune system to target cancer can be divided into active and passive immune approaches (use of approaches to induce a response or infusion of antibodies or preformed effectors), and antigen defined or undefined (peptide vaccine or tumour cell lysate) approaches, or combinations of any of these. We discuss these approaches in more details in the following sections.

Therapeutic Vaccines (Active, Antigen Specific)

Therapeutic vaccines are biological response modifiers that introduce a tumour antigen or pool of candidate antigens (or even whole tumour cells) to stimulate or restore the immune system's ability to target specific neoantigens in the tumour. These antigens are given with immunostimulant vehicles called adjuvants. In lung cancer, vaccines have been developed to target shared or tumour-specific antigens, including:

gametogenic self-antigens MAGE-3, expressed in 42% of lung cancers; NY-ESO-1, expressed in 30% of lung cancers; tumour suppressor antigen p53, mutated in 50% of lung cancers; the apoptosis regulator antigen survivin, and epithelial glycoprotein MUC1. In a phase I lung cancer clinical trial, vaccines targeting the NY-ESO-1 antigen achieved remarkable antitumour immune responses in 9 of the 10 patients treated [31]. Using a novel form of the 4-1BBL costimulatory molecule, SA-4-1BBL, as the adjuvant, a single vaccination with survivin was effective against 3LL lung tumours in mice and correlated with increased percentages of $CD8^+$ T cells expressing IFN-γ as well as potent killing responses of both $CD8^+$ T and NK cells in the absence of detectable antibodies to ssDNA as a sign of autoimmunity [32]. Many such cancer vaccines showed some promise for lung cancer in human clinical trials, but several, including MUC-1, MAGE-1 and TGF-β-transduced lung cancer cell lines, have recently been shown to be ineffective in large randomized trials [33]. Some current clinical trials of cancer vaccines in lung cancer are summarized in table 1.

Monoclonal Antibodies (Passive, Antigen Specific)

Antibodies can bind to the tumours themselves, or to substances in the blood or other tissues that help cancer cells to grow, and thus can either kill the cancer cells, inhibit their growth or keep them from spreading. Many monoclonal antibodies (mAb) are currently in use in lung cancer treatment. Two popular mAb used to treat non-squamous NSCLC include bevacizumab and cetuximab. Bevacizumab (Avastin) targets VEGF in the blood and tissues and acts to prevent the growth of new blood vessels that tumours need to grow. Bevacizumab is approved as a standard first-line treatment in combination with any platinum-based chemotherapy, followed by maintenance bevacizumab until disease progression on the basis of the US and European randomised multicentric clinical trials [34, 35]. Cetuximab (Erbitux) binds to epidermal growth factor receptor (EGFR), which is a tyrosine kinase protein, on the surface of cancer cells, thereby inhibiting their growth. The targeted delivery of drugs, such as SN-38 and doxorubicin, by an antibody has also gained interest in cancer treatment. Several other mAb that are being tested in clinical trials are included in table 2. The mAb blockade of the cytotoxic T lymphocyte antigen-4 (CTLA-4) as well as the programmed cell death-1 (PD-1) receptor and the PD-1 ligand will be discussed separately in a subsequent section of the immune checkpoint inhibition therapy.

Adoptive Cell Therapy

Another major approach of immunotherapy that is being explored for lung cancer is adoptive cell therapy, where lymphocytes from a cancer patient are genetically modified or treated with immunostimulatory reagents to enhance their activity, and then

Table 1. Therapeutic vaccines in clinical trials for lung cancer

Vaccine	Target /composition	Status	Clinical trial No.
GV1001	Targets the human telomerase reverse transcriptase, hTERT subunit of telomerase	A phase III study for patients with stage III NSCLC	NCT01579188
Tergenpumatucel-L	Consists of 3 human NSCLC cell lines genetically modified to express α-gal carbohydrates on cell surface to stimulate immune response	A phase II/III trial for NSCLC	NCT01774578
TG4010	Targets the MUC1 antigen	A phase II/III study for stage IV NSCLC	NCT01383148
DRibbles (DPV-001)	Consists of over 25 tumour antigens including 9 from the NCI's list of prioritized cancer antigens and multiple DAMPs, such as S100A8, nucleolin, calreticulin, HMGB1, HSP70, HSP90, DNAs and RNAs that bind to TLRs	A phase II trial for stage III NSCLC	NCT01909752
MUC1	Epithelial glycoprotein-based vaccine	A phase I/II trial for any stage of NSCLC	NCT01720836
CV9202	mRNA-derived NSCLC vaccine that expresses 6 tumour antigens	A phase I trial for stage IV NSCLC	NCT01915524
Wilms tumour antigen (WT1)	Recombinant WT1 peptides containing HLA-A2.1-binding anchor motifs that stimulate CTL responses	A phase II trial for mesothelioma after completing surgery and chemotherapy and/or radiation	NCT01265433
TroVax	Targets the tumour-associated antigen, 5T4, with a pox virus vector, widely expressed on solid cancers	A phase II trial for patients with mesothelioma	NCT01569919

adoptively transferred into the patient with the goal of improving the lymphocytes' cytolytic activity against cancer cells. Recent approaches use in vitro stimulation with patient-specific somatically mutant epitopes to selectively amplify tumour-specific T cell clones [ESMO immunotherapy conference, Geneva, 2015; Yang, unpubl. data]. Several clinical trials of adoptive cell transfer techniques are currently in progress in lung cancer.

In an early phase II clinical trial with stage III NSCLC, the combination of adoptive T cell therapy with IL-2 followed by standard chemotherapy and radiotherapy did not prove to be effective [36]. The reverse schedule of therapy was not tested even though previous randomized trials had demonstrated some clinical activity when chemotherapy was followed by T cell therapy [37]. However, a recent evaluation of 17 randomized controlled trials of NSCLC between 2008 and 2013, with a total of 1,172 patients, showed significant benefits of CIK cell transfers in the overall survival, objective clinical response and T lymphocyte responses without any evidence

Table 2. Monoclonal antibodies in the treatment of lung cancer

Antibody	Target	Trial status	Clinical trial No.
Bevacizumab (Avastin)	VEGF-A	Approved by the US FDA on October 11, 2006, for use as first-line therapy in advanced non-squamous NSCLC in combination with carboplatin/paclitaxel	NCT00021060 AVAPERL (MO22089)
Cetuximab (Erbitux)	EGFR	A phase II trial for stage IIIB NSCLC	NCT01059188
Bavituximab	Binds to phosphatidylserine exposed on tumour cells	Being tested with docetaxel in a phase III trial for late-stage non-squamous NSCLC	NCT01999673
Patritumab	Human epidermal growth factor receptor-3 (HER3)	Being tested in combination with erlotinib in a phase III trial for locally advanced or metastatic NSCLC	NCT02134015
Rilotumumab	Hepatocyte growth factor (HGF)	A phase II/III trial for squamous NSCLC, and a phase I/II study along with erlotinib for recurrent or progressive NSCLC	NCT02154490 NCT01233687
Cixutumumab	Insulin-like growth factor-1 receptor (IGF-1R)	A phase II trial for NSCLC	NCT01263782
IMMU-132	Antibody-drug conjugate of anti-TROP-2 and SN-38	A phase I/II study for epithelial cancers including NSCLC and SCLC	NCT01631552
Demcizumab	Notch delta-like ligand 4 (DLL4)	Being tested in combination with chemotherapy in a phase I/II trial for untreated extensive stage SCLC	NCT01859741

of major adverse events [38]. CIK cells are a group of immune effector cells featuring a mixed T and NK cell-like phenotype generated by ex vivo incubation of human peripheral blood mononuclear cells or cord blood mononuclear cells with IFN-γ, anti-CD3 and recombinant human interleukins IL-1 and IL-2. A summary of current adoptive cell transfer trials focused on T cells engineered to target various tumour antigens, including NY-ESO and MAGE, or T cells prepared from the tumour infiltrate is presented in table 3.

Adoptive transfer of non-T effector cells have also been tested in lung cancer patients. Effector γδ T cells secrete cytokines abundantly and exert potent cytotoxicity against a wide range of cancer cells. A recent phase I clinical study evaluated the safety and potential antitumour effects of γδ T cell transfers in patients with advanced or recurrent NSCLC refractory to conventional treatments. Adoptive transfer of zoledronate-expanded γδ T cells was safe, and provided a median survival of 589 days and median progression-free survival of 126 days in NSCLC patients [39].

The activity of NK cells, another potent effector cell type, has also been examined in the context of lung cancer. Some encouraging antitumour responses, with no long-term side effects, were seen in patients with advanced lung cancer following infusions with human $CD56^+$ cells from the NK-92 cell line generated from a patient with

Table 3. Current adoptive T cell transfer trials in lung cancer

T cells	Trial status	Clinical trial No.
Genetically engineered T cells targeted to:		
NY-ESO-1	A phase II trial of NY-ESO-1-specific T cells with or without dendritic cells pulsed with NY-ESO-1 antigen as a vaccine for advanced or refractory malignancies, including lung cancer	NCT01697527 NCT01967823
VEGFR2	A phase I/II trial of VEGFR2-specific T cells in aggressive cancers that have not responded to standard therapy, including lung cancer	NCT01218867
MAGE-A3	A phase I/II trial of MAGE-A3-specific T cells in metastatic cancers, including lung cancer	NCT02111850
Mesothelin	A phase I/II trial of mesothelin-specific T cells in mesothelioma	NCT01583686
T cells from the tumour infiltrate	A phase II trial of tumour-infiltrating lymphocytes in NSCL following chemotherapy	NCT02133196

lymphoma. Despite the allogeneic nature of NK-92, the cells persisted in the recipient's circulation for 48 h, with 1 out of 7 patients developing human leukocyte antigen antibodies [40]. With the administration of the IL18-IL2 fusion protein in mouse models of murine Lewis lung carcinoma, human large cell lung carcinoma (NCI-H460) enhanced IFN-γ production and NK cell cytotoxicity, resulting in significant tumour regression [41]. A study with EGFR inhibitors, including erlotinib or gefitinib, showed that these inhibitors can enhance the susceptibility of lung cancer cells A549, NCI-H23 and SW-900 to NK cell-mediated lysis by the induction of ULBP1 encoding NKG2D ligands on cancer cells and via inhibition of the protein kinase C pathway [42]. The adoptive transfer of a mixture of NK cells and cytotoxic T lymphocytes prolonged overall survival in subgroups of NSCLC males (<60 years of age, clinical stage IIIb+IV, no brain metastases, without radiotherapy, chemotherapy of >6 cycles, no application of TKI and TKI invalid) and increased their 2-year survival rate [43]. Moreover, in a metastatic lung cancer mouse model, repetitive infusions of ex vivo expanded NK cells were observed to induce antitumour T cell responses, which were further potentiated by Treg depletion [44].

Immune Checkpoint Inhibition Approaches

Our recent understanding that tumours can evade their destruction via the dysregulation of immune inhibitory checkpoint signals has led to the development of immunotherapeutic use of immune checkpoint inhibitors. These inhibitors target molecules that serve as checks and balances in the regulation of lymphocyte effector responses.

Table 4. Use of immune checkpoint inhibitors in lung cancer

Checkpoint blockade	Antibody	Trial status	Clinical trial No.
CTLA-4	Ipilimumab (Yervoy)	Phase III trials for NSCLC and SCLC	NCT01285609 NCT01450761
	Tremelimumab	Phase II clinical trial for mesothelioma	NCT01655888
PD-1	Pembrolizumab (Keytruda)	Approved by the US FDA in September 2014 and granted a 'Breakthrough Therapy' designation for the treatment of NSCLC	NCT01876511 NCT01905657 NCT02085070 NCT02054806
	Nivolumab (Opdivo)	Approved by the US FDA in March 2015 for the treatment of squamous NSCLC that has failed chemotherapy	NCT00730639 NCT02066636
PD-L1	MPDL3280A	Phase I/II trial for locally advanced or metastatic NSCLC	NCT01846416 NCT02031458 NCT01375842
	MEDI4736	Phase II/III trial for locally advanced or metastatic NSCLC, or second-line therapy in recurrent stage IIIB-IV NSCLC	NCT02154490 NCT02087423 NCT01693562 NCT02088112

By blocking inhibitory molecules or, alternatively, activating stimulatory molecules on cytolytic lymphocytes, these checkpoint treatments are designed to sustain lymphocyte antitumour responses. Immune checkpoint therapies include the monoclonal antibody blockade of CTLA-4 as well as antibody blockade of the PD-1 receptor and the PD-1 ligand. These immune checkpoint approaches have been evaluated in both NSCLC and SCLC with early evidence of activity. Recent studies with these checkpoint modulators, summarised in table 4, have shown promising results with prolonged clinical responses and tolerable toxicity in metastatic NSCLC and SCLC. The anti-PD-1 drug pembrolizumab showed a 19.4% response rate in patients with NSCLC. Moreover, some studies have demonstrated significantly improved response rates when the tumours expressed high levels of ligand PD-L1, suggesting a role for PD-L1 as a predictive immune biomarker [45]. Thus, features that are associated with good response rates are PD-L1 positivity in infiltrating immune cells as well as smoking history [46] and a high mutation burden, resulting in a higher antigenic repertoire [47].

The US FDA recently approved the use of another anti-PD-1, nivolumab, to treat patients with advanced metastatic squamous NSCLC with progression on or after platinum-based chemotherapy based on the promising results of an ongoing clinical trial [48]. To further improve the therapeutic efficacy, the immune checkpoint inhibition is also being tested actively in combination with chemotherapy or targeted therapies in locally advanced or metastatic lung cancer patients, as listed in table 5.

Table 5. Use of combination immune checkpoint treatments in lung cancer

Combination checkpoint treatments	Trial status	Clinical trial No.
Ipilimumab plus		
Chemotherapy	A phase II trial testing ipilimumab in combination with paclitaxel, cisplatin and carboplatin for NSCLC before surgery	NCT01820754
Erlotinib/crizotinib	A phase I trial for stage IV NSCLC with EGFR or ALK mutations	NCT01998126
Gleevec	A phase I trial for advanced lung cancer	NCT01738139 NCT01454102
Nivolumab	Phase I/II trials for advanced NSCLC or SCLC	NCT01928394
BMS-986015 (anti-KIR)	A phase I trial for selected metastatic and/or unresectable lung cancer	NCT01750580
Pembrolizumab	A phase I/II trial in unresectable or metastatic NSCLC	NCT02039674
Tremelimumab plus		
MEDI4736	Phase I/Ib trials for advanced NSCLC	NCT01975831 NCT02000947
Nivolumab plus		
Azacitidine/Entinostat	A phase II trial testing nivolumab given after epigenetic priming drugs in NSCLC	NCT01928576
Chemotherapy/targeted therapy	A phase I trial testing nivolumab with gemcitabine/cisplatin, pemetrexed/cisplatin, carboplatin/paclitaxel, bevacizumab maintenance and erlotinib in stage IIIB/IV NSCLC	NCT01454102
Lirilumab (anti-KIR)	A phase I trial for metastatic and/or unresectable lung cancer	NCT01714739
BMS-986016 (anti-LAG3)	A phase I trial for lung cancer	NCT01968109
Pembrolizumab plus		
Chemotherapy/targeted therapy	A phase I/II trial testing pembrolizumab with Alimta, bevacizumab, carboplatin, Iressa, Taxol and erlotinib in unresectable or metastatic NSCLC	NCT02039674
IDO inhibitors	A phase I/II trial testing pembrolizumab with MK-3475 and INCB024360 in advanced NSCLC	NCT02178722
MPDL3280A plus		
Bevacizumab/chemotherapy	A phase Ib trial testing combinatorial Abraxane, Alimta, carboplatin, Eloxatin, fluorouracil, Taxol, leucovorin for advanced or metastatic NSCL	NCT01633970
MEDI4736 plus		
Gefitinib (EGFR inhibitor)	A phase I trial for NSCL	NCT02088112

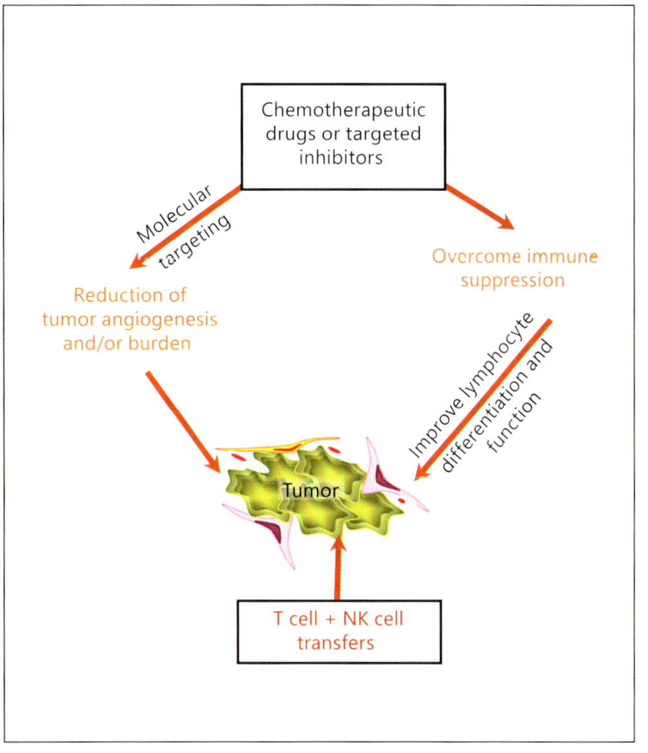

Fig. 1. A putative immunotherapeutic clinical protocol capable of targeting a broad spectrum of cancer cells. A synergistic combinatorial regimen involving adoptive transfer of NK cells and T lymphocytes to control metastatic antigen-escape variants of tumour cells would incorporate complementary immunostimulatory agents that can overcome the tumour-associated immunosuppression and chemotherapeutic drugs or targeted inhibitors, which can reduce tumour angiogenesis and/or burden. Such carefully optimized, multiple components of combinatorial immunotherapeutic approaches would be necessary for the effective eradication of cancer and relapse-free survival.

Conclusions and Future Perspectives

The immune checkpoint PD-1 antibodies have shown superior responses over CTLA-4 antibodies, most likely because the PD-1 pathway blockade acts locally in the tumour microenvironment. Also, the response of lung cancer to immune checkpoint inhibition appears to be independent of driver mutations present in most lung cancers. However, this needs to be investigated more carefully as mutational load in cancer cells may give rise to neoepitopes that may facilitate immune recognition by effector lymphocytes. Indeed, in a recent study a higher nonsynonymous mutation burden in NSCLC patients treated with pembrolizumab correlated with the higher neoantigen repertoire, which was associated with an improved objective response, durable clinical benefit and progression-free survival in one responder [47]. Thus, the genomic landscape of lung cancers may shape the response to immune checkpoint therapy.

Moreover, treatment with the PD-1 inhibitors results in an objective durable response only in a minority of patients. Further research is needed to find a reliable biomarker of response in order to guarantee that only patients who will respond receive treatment. However, due to the dynamic nature of immune response to tumours and the complexity of the multiple immune checkpoint pathways, it will be important to identify synergistic combinatorial regimens that could target a broad spectrum of cancer cells. An emerging body of evidence in various models of immune rejection, including cancer, indicates an indispensable cross-talk between innate effector cells and T lymphocytes that results in an optimal effector response capable of rejecting metastatic antigen-escape variants of tumour cells [49–52]. While some patients may have a durable benefit from PD-1 pathway inhibition alone, an optimal clinical protocol may combine these antitumour strategies with complementary immunostimulatory agents that can overcome the tumour-associated immunosuppression, such as anti-CTLA4, bevacizumab, adenosine receptor antagonists, or stimulation of Notch signalling in lymphopoietic cells by immunopotentiating Notch ligand formulations [14]. Given the heterogenous nature of rapidly evolving cancer cells, such carefully optimized, multiple components of combinatorial immunotherapeutic approaches would be necessary for the effective eradication of cancer and relapse-free survival in the largest number of patients.

Acknowledgments

We thank Ariana Renrick for help in the literature search and preparation of the tables.

This work was funded by National Institutes of Health grants R01CA138923 (M.M.D.), RO1CA175370 (M.M.D. and D.P.C.), P50CA90949 (M.M.D. and A.S.), U54 MD007593 (A.S.), U54 CA163069 (A.S.) and SC1 CA182843 (A.S.), as well as an Ohio State University Drug Development Institute grant (M.M.D.), and Dallapezze Fund (M.M.D. and D.P.C.).

References

1 Rakoff-Nahoum S, Medzhitov R: Toll-like receptors and cancer. Nat Rev Cancer 2009;9:57–63.

2 Massarelli E, Papadimitrakopoulou V, Welsh J, Tang C, Tsao AS: Immunotherapy in lung cancer. Transl Lung Cancer Res 2014;3:53–63.

3 Cekic C, Day YJ, Sag D, Linden J: Myeloid expression of adenosine A2A receptor suppresses T and NK cell responses in the solid tumor microenvironment. Cancer Res 2014;74:7250–7259.

4 Tsartsalis D, Grapsa D, Skopeliti M, Dragioti E, Charpidou A, Politi E, Tsitsilonis O, Syrigos K: In vitro exposure of NK-92 cells to serum from patients with non-small cell lung cancer impairs their cytotoxicity. Anticancer Res 2015;35:1543–1548.

5 Jin S, Deng Y, Hao JW, Li Y, Liu B, Yu Y, Shi FD, Zhou QH: NK cell phenotypic modulation in lung cancer environment. PloS One 2014;9:e109976.

6 Pircher A, Gamerith G, Amann A, Reinold S, Popper H, Gachter A, Pall G, Woll E, Jamnig H, Gastl G, Wolf AM, Hilbe W, Wolf D: Neoadjuvant chemo-immunotherapy modifies $CD4^{+}CD25^{+}$ regulatory T cells (Treg) in non-small cell lung cancer (NSCLC) patients. Lung Cancer 2014;85:81–87.

7 Hodge G, Barnawi J, Jurisevic C, Moffat D, Holmes M, Reynolds PN, Jersmann H, Hodge S: Lung cancer is associated with decreased expression of perforin, granzyme B and interferon (IFN)-γ by infiltrating lung tissue T cells, natural killer (NK) T-like and NK cells. Clin Exp Immunol 2014;178:79–85.

8 Donatelli SS, Zhou JM, Gilvary DL, Eksioglu EA, Chen X, Cress WD, Haura EB, Schabath MB, Coppola D, Wei S, Djeu JY: TGF-β-inducible microRNA-183 silences tumor-associated natural killer cells. Proc Natl Acad Sci U S A 2014;111:4203–4208.

9 Torre-Amione G, Beauchamp RD, Koeppen H, Park BH, Schreiber H, Moses HL, Rowley DA: A highly immunogenic tumor transfected with a murine transforming growth factor type β_1 cDNA escapes immune surveillance. Proc Natl Acad Sci U S A 1990; 87:1486–1490.

10 Kim R, Emi M, Tanabe K: Functional roles of immature dendritic cells in impaired immunity of solid tumour and their targeted strategies for provoking tumour immunity. Clin Exp Immunol 2006;146: 189–196.

11 Wang Y, Han G, Wang K, Liu G, Wang R, Xiao H, Li X, Hou C, Shen B, Guo R, Li Y, Chen G: Tumor-derived GM-CSF promotes inflammatory colon carcinogenesis via stimulating epithelial release of VEGF. Cancer Res 2014;74:716–726.
12 Zhao P, Bu X, Wei X, Sun W, Xie X, Li C, Guo Q, Zhu D, Gao D: Dendritic cell immunotherapy combined with cytokine-induced killer cells promotes skewing toward Th2 cytokine profile in patients with metastatic non-small cell lung cancer. Int Immunopharmacol 2015;25:450–456.
13 Huang Y, Chen X, Dikov MM, Novitskiy SV, Mosse CA, Yang L, Carbone DP: Distinct roles of VEGFR-1 and VEGFR-2 in the aberrant hematopoiesis associated with elevated levels of VEGF. Blood 2007;110: 624–631.
14 Huang Y, Lin L, Shanker A, Malhotra A, Yang L, Dikov MM, Carbone DP: Resuscitating cancer immunosurveillance: selective stimulation of DLL1-Notch signaling in T cells rescues T-cell function and inhibits tumor growth. Cancer Res 2011;71:6122–6131.
15 Ohm JE, Gabrilovich DI, Sempowski GD, Kisseleva E, Parman KS, Nadaf S, Carbone DP: VEGF inhibits T-cell development and may contribute to tumor-induced immune suppression. Blood 2003;101: 4878–4886.
16 Ryzhov S, Biktasova A, Goldstein AE, Zhang Q, Biaggioni I, Dikov MM, Feoktistov I: Role of JunB in adenosine A2B receptor-mediated vascular endothelial growth factor production. Mol Pharmacol 2014; 85:62–73.
17 Ryzhov S, Novitskiy SV, Zaynagetdinov R, Goldstein AE, Carbone DP, Biaggioni I, Dikov MM, Feoktistov I: Host A_{2B} adenosine receptors promote carcinoma growth. Neoplasia 2008;10:987–995.
18 Ryzhov S, Novitskiy SV, Goldstein AE, Biktasova A, Blackburn MR, Biaggioni I, Dikov MM, Feoktistov I: Adenosinergic regulation of the expansion and immunosuppressive activity of CD11b+Gr1+ cells. J Immunol 2011;187:6120–6129.
19 Visco C, Vassilakopoulos TP, Kliche KO, Nadali G, Viviani S, Bonfante V, Medeiros V, Notti P, Rassidakis GZ, Peethambaram P, Wilder R, Witzig T, Gianni M, Bonadonna G, Pizzolo G, Pangalis GA, Cabanillas F, Sarris AH: Elevated serum levels of IL-10 are associated with inferior progression-free survival in patients with Hodgkin's disease treated with radiotherapy. Leuk Lymphoma 2004;45:2085–2092.
20 Fayad L, Keating MJ, Reuben JM, O'Brien S, Lee BN, Lerner S, Kurzrock R: Interleukin-6 and interleukin-10 levels in chronic lymphocytic leukemia: correlation with phenotypic characteristics and outcome. Blood 2001;97:256–263.
21 Kim J, Modlin RL, Moy RL, Dubinett SM, McHugh T, Nickoloff BJ, Uyemura K: IL-10 production in cutaneous basal and squamous cell carcinomas: a mechanism for evading the local T cell immune response. J Immunol 1995;155:2240–2247.
22 Link AA, Kino T, Worth JA, McGuire JL, Crane ML, Chrousos GP, Wilder RL, Elenkov IJ: Ligand-activation of the adenosine A2a receptors inhibits IL-12 production by human monocytes. J Immunol 2000; 164:436–442.
23 Rossowska J, Anger N, Kicielinska J, Pajtasz-Piasecka E, Bielawska-Pohl A, Wojas-Turek J, Dus D: Temporary elimination of IL-10 enhanced the effectiveness of cyclophosphamide and BMDC-based therapy by decrease of the suppressor activity of MDSCs and activation of antitumour immune response. Immunobiology 2015;220:389–398.
24 Steinbrink K, Graulich E, Kubsch S, Knop J, Enk AH: $CD4^+$ and $CD8^+$ anergic T cells induced by interleukin-10-treated human dendritic cells display antigen-specific suppressor activity. Blood 2002;99: 2468–2476.
25 Terabe M, Matsui S, Noben-Trauth N, Chen H, Watson C, Donaldson DD, Carbone DP, Paul WE, Berzofsky JA: NKT cell-mediated repression of tumor immunosurveillance by IL-13 and the IL-4R-STAT6 pathway. Nat Immunol 2000;1:515–520.
26 Curiel TJ, Coukos G, Zou L, Alvarez X, Cheng P, Mottram P, Evdemon-Hogan M, Conejo-Garcia JR, Zhang L, Burow M, Zhu Y, Wei S, Kryczek I, Daniel B, Gordon A, Myers L, Lackner A, Disis ML, Knutson KL, Chen L, Zou W: Specific recruitment of regulatory T cells in ovarian carcinoma fosters immune privilege and predicts reduced survival. Nat Med 2004;10:942–949.
27 Ishida T, Ishii T, Inagaki A, Yano H, Komatsu H, Iida S, Inagaki H, Ueda R: Specific recruitment of CC chemokine receptor 4-positive regulatory T cells in Hodgkin lymphoma fosters immune privilege. Cancer Res 2006;66:5716–5722.
28 Uyttenhove C, Pilotte L, Theate I, Stroobant V, Colau D, Parmentier N, Boon T, van den Eynde BJ: Evidence for a tumoral immune resistance mechanism based on tryptophan degradation by indoleamine 2,3-dioxygenase. Nat Med 2003;9:1269–1274.
29 Lee GK, Park HJ, Macleod M, Chandler P, Munn DH, Mellor AL: Tryptophan deprivation sensitizes activated T cells to apoptosis prior to cell division. Immunology 2002;107:452–460.
30 Friberg M, Jennings R, Alsarraj M, Dessureault S, Cantor A, Extermann M, Mellor AL, Munn DH, Antonia SJ: Indoleamine 2,3-dioxygenase contributes to tumor cell evasion of T cell-mediated rejection. Int J Cancer 2002;101:151–155.

31 Kakimi K, Isobe M, Uenaka A, Wada H, Sato E, Doki Y, Nakajima J, Seto Y, Yamatsuji T, Naomoto Y, Shiraishi K, Takigawa N, Kiura K, Tsuji K, Iwatsuki K, Oka M, Pan L, Hoffman EW, Old LJ, Nakayama E: A phase I study of vaccination with NY-ESO-1f peptide mixed with Picibanil OK-432 and Montanide ISA-51 in patients with cancers expressing the NY-ESO-1 antigen. Int J Cancer 2011;129:2836–2846.

32 Srivastava AK, Sharma RK, Yolcu ES, Ulker V, MacLeod K, Dinc G, Shirwan H: Prime-boost vaccination with SA-4-1BBL costimulatory molecule and survivin eradicates lung carcinoma in CD8+ T and NK cell dependent manner. PLoS One 2012;7: e48463.

33 Kelly RJ, Giaccone G: Lung cancer vaccines. Cancer J 2011;17:302–308.

34 Barlesi F, Gervais R, Lena H, Hureaux J, Berard H, Paillotin D, Bota S, Monnet I, Chajara A, Robinet G: Pemetrexed and cisplatin as first-line chemotherapy for advanced non-small-cell lung cancer (NSCLC) with asymptomatic inoperable brain metastases: a multicenter phase II trial (GFPC 07-01). Ann Oncol 2011;22:2466–2470.

35 Reck M, von Pawel J, Zatloukal P, Ramlau R, Gorbounova V, Hirsh V, Leighl N, Mezger J, Archer V, Moore N, Manegold C: Overall survival with cisplatin-gemcitabine and bevacizumab or placebo as first-line therapy for nonsquamous non-small-cell lung cancer: results from a randomised phase III trial (AVAiL). Ann Oncol 2010;21:1804–1809.

36 Ratto GB, Cafferata MA, Scolaro T, Bruzzi P, Alloisio A, Costa R, Spessa E, Semino C, Melioli G: Phase II study of combined immunotherapy, chemotherapy, and radiotherapy in the postoperative treatment of advanced non-small-cell lung cancer. J Immunother 2000;23:161–167.

37 Ratto GB, Zino P, Mirabelli S, Minuti P, Aquilina R, Fantino G, Spessa E, Ponte M, Bruzzi P, Melioli G: A randomized trial of adoptive immunotherapy with tumor-infiltrating lymphocytes and interleukin-2 versus standard therapy in the postoperative treatment of resected nonsmall cell lung carcinoma. Cancer 1996;78:244–251.

38 Wang M, Cao JX, Pan JH, Liu YS, Xu BL, Li D, Zhang XY, Li JL, Liu JL, Wang HB, Wang ZX: Adoptive immunotherapy of cytokine-induced killer cell therapy in the treatment of non-small cell lung cancer. PLoS One 2014;9:e112662.

39 Kakimi K, Matsushita H, Murakawa T, Nakajima J: Gammadelta T cell therapy for the treatment of non-small cell lung cancer. Transl Lung Cancer Res 2014; 3:23–33.

40 Tonn T, Schwabe D, Klingemann HG, Becker S, Esser R, Koehl U, Suttorp M, Seifried E, Ottmann OG, Bug G: Treatment of patients with advanced cancer with the natural killer cell line NK-92. Cytotherapy 2013;15:1563–1570.

41 Du G, Ye L, Zhang G, Dong Q, Liu K, Tian J: Human IL18–IL2 fusion protein as a potential antitumor reagent by enhancing NK cell cytotoxicity and IFN-γ production. J Cancer Res Clin Oncol 2012;138: 1727–1736.

42 Kim H, Kim SH, Kim MJ, Kim SJ, Park SJ, Chung JS, Bae JH, Kang CD: EGFR inhibitors enhanced the susceptibility to NK cell-mediated lysis of lung cancer cells. J Immunother 2011;34:372–381.

43 Zhang G, Zhao H, Wu J, Li J, Xiang Y, Wang G, Wu L, Jiao S: Adoptive immunotherapy for non-small cell lung cancer by NK and cytotoxic T lymphocytes mixed effector cells: retrospective clinical observation. Int Immunopharmacol 2014;21:396–405.

44 Salagianni M, Baxevanis CN, Papamichail M, Perez SA: New insights into the role of NK cells in cancer immunotherapy. Oncoimmunology 2012;1:205–207.

45 Garon EB, Rizvi NA, Hui R, Leighl N, Balmanoukian AS, Eder JP, Patnaik A, Aggarwal C, Gubens M, Horn L, Carcereny E, Ahn MJ, Felip E, Lee JS, Hellmann MD, Hamid O, Goldman JW, Soria JC, Dolled-Filhart M, Rutledge RZ, Zhang J, Lunceford JK, Rangwala R, Lubiniecki GM, Roach C, Emancipator K, Gandhi L, Pembrolizumab for the treatment of non-small-cell lung cancer. New Engl J Med 2015; 372:2018–2028.

46 Herbst RS, Soria JC, Kowanetz M, Fine GD, Hamid O, Gordon MS, Sosman JA, McDermott DF, Powderly JD, Gettinger SN, Kohrt HE, Horn L, Lawrence DP, Rost S, Leabman M, Xiao Y, Mokatrin A, Koeppen H, Hegde PS, Mellman I, Chen DS, Hodi FS: Predictive correlates of response to the anti-PD-L1 antibody MPDL3280A in cancer patients. Nature 2014; 515:563–567.

47 Rizvi NA, Hellmann MD, Snyder A, Kvistborg P, Makarov V, Havel JJ, Lee W, Yuan J, Wong P, Ho TS, Miller ML, Rekhtman N, Moreira AL, Ibrahim F, Bruggeman C, Gasmi B, Zappasodi R, Maeda Y, Sander C, Garon EB, Merghoub T, Wolchok JD, Schumacher TN, Chan TA: Cancer immunology: mutational landscape determines sensitivity to PD-1 blockade in non-small cell lung cancer. Science 2015; 348:124–128.

48 Gettinger SN, Horn L, Gandhi L, Spigel DR, Antonia SJ, Rizvi NA, Powderly JD, Heist RS, Carvajal RD, Jackman DM, Sequist LV, Smith DC, Leming P, Carbone DP, Pinder-Schenck MC, Topalian SL, Hodi FS, Sosman JA, Sznol M, McDermott DF, Pardoll DM, Sankar V, Ahlers CM, Salvati M, Wigginton JM, Hellmann MD, Kollia GD, Gupta AK, Brahmer JR: Overall survival and long-term safety of nivolumab (anti-programmed death 1 antibody, BMS-936558, ONO-4538) in patients with previously treated advanced non-small-cell lung cancer. J Clin Oncol 2015;33:2004–2012.

49 Shanker A, Buferne M, Schmitt-Verhulst AM: Cooperative action of CD8 T lymphocytes and natural killer cells controls tumour growth under conditions of restricted T-cell receptor diversity. Immunology 2010;129:41–54.
50 Shanker A, Marincola FM: Cooperativity of adaptive and innate immunity: implications for cancer therapy. Cancer Immunol Immunother 2011;60:1061–1074.
51 Shanker A, Verdeil G, Buferne M, Inderberg-Suso EM, Puthier D, Joly F, Nguyen C, Leserman L, Auphan-Anezin N, Schmitt-Verhulst AM: CD8 T cell help for innate antitumor immunity. J Immunol 2007;179:6651–6662.
52 Malhotra A, Shanker A: NK cells: immune cross-talk and therapeutic implications. Immunotherapy 2011;3:1143–1166.

Anil Shanker, PhD
Meharry Medical College School of Medicine, 2005 Harold D. West Basic Sciences Building
1005 Dr. DB Todd Jr. Boulevard
Nashville, TN 37208 (USA)
E-Mail ashanker@mmc.edu

David P. Carbone, MD, PhD
Ohio State University Medical Center
460 W 12th Avenue, 484 BRT
Columbus, OH 43210 (USA)
E-Mail David.Carbone@osumc.edu

Michielin O, Coukos G (eds): Immuno-Oncology.
Prog Tumor Res. Basel, Karger, 2015, vol 42, pp 110–135 (DOI: 10.1159/000437180)

T Cell Engineering

Magdalena Pircher[a] · Thomas Schirrmann[b] · Ulf Petrausch[a]

[a]Department of Oncology, University Hospital Zurich, Zurich, Switzerland; [b]Institute of Biochemistry, Biotechnology and Bioinformatics, Technische Universität Braunschweig, Braunschweig, Germany

Abstract

T cells are a new and promising antigen-specific therapeutic option for the treatment of malignant diseases. To achieve antigen specificity against tumor antigens, T cells can be manipulated by gene transfer to express chimeric antigen receptors (CARs). CAR-expressing T cells are called redirected T cells. CARs are composed of an extracellular antibody-derived antigen recognition domain, a transmembrane domain and a cytoplasmatic signal domain. Therefore, redirected T cells combine the exchangeable specificity of an antibody with the cytotoxic machinery of a T cell. Early clinical trials with redirected T cells targeting cluster of differentiation (CD) 19 have shown impressive results in CD19-positive hematological cancers. However, for solid cancers only limited clinical experience exists and new and innovative concepts have to be developed to overcome tumor-mediated immune suppression. Herein, we describe the general design of a CAR, the function of the different domains and the different strategies to produce redirected T cells. Furthermore, we summarize and discuss the preclinical and clinical data indicating the tremendous potential of redirected T cells to become a mainstay of cancer immunotherapy.

T cells are a new antigen-specific therapeutic option to overcome certain limitations of monoclonal antibodies because they are capable of actively migrating into malignant tumors and directly killing cancer cells. Monoclonal antibodies are often rapidly removed from the tumor cell surface by processes like capping, shedding or endocytosis. Additionally, there is still the large hurdle of inefficient access of macromolecule-like antibodies to poorly vascularized, malignant tumors, which are additionally protected by endothelia, surrounding tissue or the blood-tissue barriers. This results in limitations regarding efficacy and clinical benefit. Even though monoclonal antibodies have revolutionized therapy of certain malignant diseases (e.g. lymphoma or breast cancer) [1] via direct antiproliferative effects by blocking signaling pathways or via the induction of direct cytotoxic effects by the recruitment of complement or Fc receptor-bearing immune cells (antibody-dependent cell-mediated cytotoxicity), nowadays new immunological therapeutic strategies such as genetically modified T cells are being tested to overcome the limitations mentioned above.

Over the last decades multiple lines of experimental data have demonstrated that T cells are involved in the tumor immune surveillance postulated by Burnet [2] and Smyth et al. [3]. T cells are able to kill tumor cells, but they are also suppressed by the tumor microenvironment [4]. To date, only a limited number of T cell receptors (TCRs) targeting antigens expressed on cancer cells are known. Since TCRs only recognize targets in the context of major histocompatibility complexes, the TCR binding is a highly individual process that is difficult to translate into a standardized, therapeutic intervention with a defined antigen-specific T cell population.

To overcome the limitations of antibody and T cell therapies the antigen specificity of antibodies should be combined with the effector properties of cytotoxic T cells. The antibody allows the targeting of a well-defined antigen, whereas the T cell offers direct target cell killing and active migration into malignant tissues. With the help of an antibody, surface antigens can be targeted independently of the major histocompatibility complex. Thereby, T cell therapy becomes broadened to a wider spectrum of antigens independent of the endogenous T cell repertoire. Additionally, the effect can be amplified by T cell proliferation, resulting in an expanded antigen-specific effector cell population. Today, there are two major strategies to redirect T cells towards malignant tumors with antibodies.

First, bispecific antibodies binding to a tumor-associated antigen as well as to signal molecules like CD3 on T cells have been successfully used to specifically redirect and activate T cells against tumor cells [5, 6]. Although bispecific antibodies can extend the natural antibody-mediated cellular cytotoxicity, they suffer from similar limitations as monospecific antibodies, like the need for large amounts of recombinant protein and their inefficient migration into solid tumors [7]. T cells activated by bispecific antibodies can lose their signal transduction and effector properties following target cell recognition [8] and require additional costimulatory signals for full activation [9]. Nevertheless, the CD3 × CD19 bispecific antibody (blinatumomab) first showed activity in phase I and phase II clinical trials for patients with follicular lymphoma, mantle-cell lymphoma and acute lymphoblastic leukemia, and achieved a breakthrough therapy designation by the US Food and Drug Administration (FDA) [10–12].

Second, genetically engineered T cells are redirected against malignant tumors via the expression of a CAR. This approach is discussed later in this chapter. T cells which are genetically engineered by antigen-specific chimeric receptors contain genes consisting of an antigen recognition domain of an antibody and signal domains triggering their cytolytic mechanisms. Importantly, the CAR can be designed to deliver additional costimulatory signals resulting in more efficient and sustained T cell activation and target cell killing. T cells expressing a CAR are called redirected or genetically modified T cells [13]. In this chapter we will focus on the mechanisms and the design of single-chain CARs which have led to the recently published breakthroughs in the development of redirected T cells, pushing the field of adoptive T cell transfer into a new promising era. Specific problems when TCRs are used to redirect T cells will not be covered here. However, we will show and discuss the design and the most important recent experimental and clinical evidence for T cells redirected by CARs.

Chimeric Antigen Receptors

Based on the most recent published data, Rosenberg and colleagues [14] foresee T cell therapy as the therapeutic intervention that can potentially achieve a cure in at least lymphatic leukemia: 'Although the palliation of cancer is the daily task of the oncologist, its cure is our "fervent hope". It seems clear that drug-based treatments alone generally do not kill all cancer cells, with the notable exception of germ cell tumors and some hematological malignancies. Residual disease after drug therapy will ultimately grow back, with lethal consequences. However, the immune system is capable of achieving sterilizing immunity and inducing long-term, durable responses that are probably curative. The use of adoptive T cell-based therapies to eradicate cancer is at a rare nexus of basic immunology and clinically meaningful therapy.'

What Is a Chimeric Antigen Receptor?

The general design of a CAR consists of an amino-terminal signal peptide, followed by an extracellular antibody-derived antigen recognition domain, a transmembrane domain and a cytoplasmatic signal domain (fig. 1). These domains are required to enable the cell surface expression of the receptor and to couple the antibody-mediated antigen recognition with the effector function of the gene-modified T cell. Additional domains, in particular extracellular spacer domains between the antigen recognition domain and transmembrane domain, can dramatically improve the expression, antigen binding and signal function of the CAR. Novel CAR designs contain additional motifs of costimulatory receptors or utilize downstream signal molecules to enhance the effector functions of the gene-modified immune cells [15].

How Is a Chimeric Antigen Receptor Built?

During the first attempts of T cell gene modification, chimeric TCR variants were designed by the substitution of both TCR variable (V) regions with those of antibodies [16, 17]. These chimeric TCR chains also interacted with endogenous TCR chains, often resulting in nonfunctional TCR hybrid complexes. The so-called 'two-chain CARs' required the transfection of two genes into the same cell resulting in inefficient

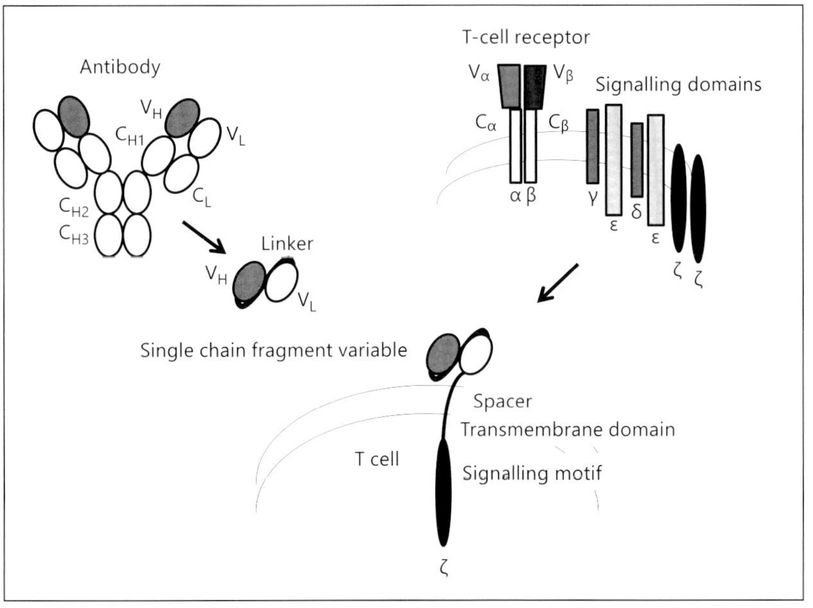

Fig. 1. CAR T cells can be grafted with antibody specificity by the transfer of a gene coding for a CAR. The employment of a scFv cloned in line with a spacer, transmembrane domain and a signalling motif allows the expression of a functional single-chain CAR. V_H = Variable heavy chain; V_L = variable light chain; $C_{H1\text{-}3}$ = constant heavy chains; C_L = constant light chain; $V_{\alpha\&\beta}$ = variable chain; $C_{\alpha\&\beta}$ = constant chain of TCRs.

and variable production of gene-modified T cells. The fusion of antibody V regions to one of the TCR signal chains like the CD3ε or ζ chain or the FcεRI γ chain allowed the construction of the first single-chain CAR. The fusion of the extracellular domain of CD4 to cytoplasmatic domains of the ζ chain resulted in a functional chimeric single chain receptor gene which could be expressed in T cells [18]. These early redirected lymphocytes demonstrated specific cytotoxicity against HIV (human immunodeficiency virus) [18]. Major interest lies in these early HIV-specific redirected T cells since the early data are now used to estimate long-term risks of redirected T cell transfer [19]. The use of Fab fragments fused to the CD3ε, ζ or FcεRI γ chain enhanced the formation of a functional antigen binding domain stabilized by the disulphide bridge between the C_L and C_{H1} domain and reduced the interaction with endogenous TCR signal chains [20, 21].

The use of single-chain fragment variable (scFv) antibody fragments allowed CAR constructs consisting of one polypeptide chain [22]. scFv are composed of the antibody V domains joined by a flexible peptide linker (fig. 1). scFv can be directly produced by phage display from human antibody gene libraries, whereas V regions derived from hybridoma antibodies must be genetically engineered and do not necessarily result in functional scFv fragments after fusion [23]. Since transcription and translation of one polypeptide chain results in the efficient expression of a functional CAR, only single-chain CARs have been used in recent clinical trials [24].

Chimeric Antigen Receptor Signaling – From Coach to Race Wagon
As for TCR-mediated T cell activation, the 'two-signal' paradigm is also valid for CAR-mediated T cell activation. As the first signal domain is mostly the ζ chain of the TCR complex and the FcεRI γ chain, two disulphide-linked homodimers with a very short extraplasmatic domain have been used for CARs. Their signal transduction is mainly mediated by cytoplasmatic activatory sequence motifs containing two tyrosine phosphorylation sites. These immune-receptor tyrosine activation motifs recruit nonreceptor protein tyrosine kinases (PTKs) of the Src family, such as Syk and ZAP70 (ζ-associated protein 70) via their SH2 (Src homology 2) domains. The PTKs activate a multitude of signal pathways [25]. The individual immune-receptor tyrosine activation motifs of the ζ chain are recognized by different downstream signaling molecules [26], suggesting not only multiple signaling events, but also a strong regulation. The positive selection and T cell maturation is more efficiently supported by the ζ chain than by the γ chain [27], whereas both signal chains seem to equally mediate T cell development and function. T cells transfected with CAR gene constructs containing the ζ chain instead of the FcεRI γ chain showed a greater capacity of tumor control [28]. However, binding of the CAR and phosphorylation of the receptor alone is not sufficient to obtain an efficient T cell activation resulting in tumor cell killing [29, 30]. As proposed by the 'two-signal' paradigm, T cells require a second and optionally a third signal to induce activation and target cell killing after phosphorylation of the ζ chain of the CAR. Multiple lines of experimentation indicated that an appropriate second signal chain must be included into the CAR construct or the receptor has to be assembled with cellular signal molecules triggering the cytotoxic mechanisms of the effector cell [31–35]. Therefore, the role of the second and third signal after activation of the T cell through the CAR has recently gained more interest. It became obvious that different secondary and third signals are important for different T cell subsets with different functions [15]. CARs were designed to provide further activation signals as part of the polypeptide chain. Based on the number of costimulatory signals, CARs are now classified in three generations (fig. 2).

The first generation of CARs was composed of the scFv, extracellular spacer, transmembrane domain and the ζ chain of the TCR complex or the FcεRI γ chain. To achieve the second signal different costimulatory receptors were tested. The introduction of CD28 signal domains achieved costimulatory properties. First, as proof of principle, double transfectants with scFv-CD28 and scFv-ζ showed synergistic signal effects [36]. Second-generation CARs introduced signal domains of CD28 and ζ chain into the one receptor construct. These CAR constructs mediated both cytotoxic and costimulatory signals in T cells [37, 38]. Nevertheless, there were also controversial

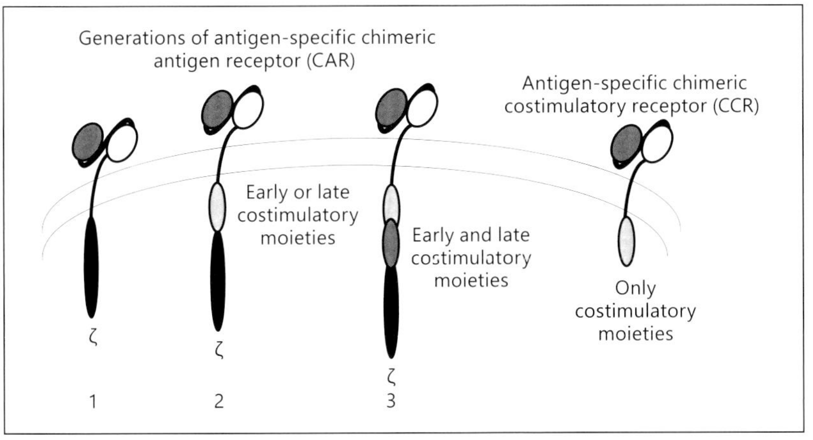

Fig. 2. Generation of CARs. The first generation of CARs comprised the scFv, extracellular spacer, transmembrane domain and the ζ chain of the TCR complex (or the FcεRI γ chain). To provide the needed costimulation additional signaling domains were integrated into the CAR. 'Early' costimulatory moieties like CD28 and 'late' signals like CD134 (OX40) or 4-1BB were used in the second generation of CARs. The combination of early and late costimulation resulted in third-generation CARs. CCRs were also designed to provide costimulation only.

results regarding the requirement of CD28 costimulation of CAR-transfected T cells [34, 39]. The main effect of CD28 signaling is the secretion of interleukin (IL)-2, which serves as the autocrine and paracrine survival factor. On the flip side, IL-2 can maintain regulatory T cells in the tumor microenvironment and thereby block antitumor immunity [40]. During T cell activation CD28 is the first costimulatory molecule and is downregulated after activation. The third generation of CARs includes two costimulatory signals, one 'early', such as CD28, and one 'late', such as CD134 (OX40) or 4-1BB. 'Late' costimulatory molecules of the tumor necrosis factor receptor family result in enhanced proliferation and survival of T cells. OX40 is upregulated in T cells 48 h after activation [41]. Initially OX40 is not required for T cell activation. However, preclinical models showed that OX40-deficient T cells proliferate less than 12 days after activation [42]. Therefore, a third survival signal for the redirected T cell seems to be important. The introduction of CD28 and CD134 domains into one CAR construct resulted in activation, proliferation and elevated survival signals [43]. 4-1BB is another costimulatory molecule providing survival signals through PI3 kinase/Akt activation. 4-1BB and CD28 CARs led to a prolonged survival of redirected T cells [44, 45]. However, there are also CAR constructs designed as second-generation CARs providing only 'late' costimulatory signals. 4-1BB as only the secondary signal in a CAR directed against CD19 caused clinical responses and formation of long-term memory T cells [46, 47]. To further modulate the functionality of redirected T cells, antigen-specific chimeric costimulatory receptors (CCRs) were designed and investigated. These CCRs only provide a costimulatory signal when the scFv binds to the target. This approach is intended to target 'antigen pairs' exclusively expressed on malignant cells and not on healthy ones. The CAR binds to one antigen (signal one) and the CCR binds to a second antigen on the same cell (signal two). Only both signals induce a sufficient T cell response [48, 49].

The Transmembrane Domain – More Than an Anchor

Besides its function as a membrane anchor, the transmembrane domain of the CAR construct plays an important role in the association with other signal components (fig. 1). It has been noted that transmembrane moieties of CD3 components may negatively affect CAR responses [50]; however, there are also results showing that the exchange of the transmembrane domain did not affect the signal transduction capacity of the ζ chain [51]. There are studies describing the effects of transmembrane domains chosen from CD4, CD28 and CD8 in CAR constructs [18, 43, 52, 53]. For example, the transmembrane domain of CD8α was crucial for CAR-containing cytoplasmatic nonreceptor PTKs as the signal domain, probably caused by different assembly and clustering of the chimeric receptor [54, 55]. Therefore, the choice of the optimal transmembrane domain may depend on the respective CAR construct.

The Extracellular Spacer – The Functionalizer

Generally, chimeric receptor constructs composed of the ζ chain take more advantage from spacer domains than receptor constructs containing the γ chain [56]. This is probably due to the stabilization of the CAR preventing an intracellular proteolytic breakdown of the ζ chain and, therefore, resulting in a stable membrane expression. The spacer is introduced in the majority of CAR constructs between the scFv fragment and the transmembrane moiety, and its main function is to increase the distance between the antigen recognition domain and the plasma membrane (fig. 1). Furthermore, hinge-like properties improved the accessibility of the antigen-binding region to the antigen [57].

Several investigators used the human IgG1 Fc part (hinge-CH_2-CH_3 domains) as a spacer domain into the CAR constructs [58, 59]. The Fc part also promotes and stabilizes the CAR homodimerization and suppresses an interaction with the endogenous ζ chain [60]. However, the Fc part can be recognized by NK cells leading to an unspecific inflammation. Therefore, the Fc part can be mutated to prevent NK cell activation and off-target toxicity [61]. Other groups employed the D3/D4 Ig-like domains of CD4, the hinge region of CD8α [47, 57, 62] and domains of CD28 [63, 64].

Gene Transfer – The Redirection

Most studies used retroviral or lentiviral systems for the gene transfer of CARs into T cells. Optimized retroviral gene delivery systems allow the transduction of primary T cells with a high efficiency [65–67]. Most of these systems are based on the Moloney murine leukemia virus (MoMuLV), the amphotropic murine leukemia virus. The pseudotyping of the viral surface of MoMuLV vectors with the envelope of amphotropic murine leukemia virus or gibbon ape leukemia virus significantly improved the transfection efficiencies of human T cells [68–70]. Coculturing of primary T cells with the retrovirus-producing packaging cell line resulted in transduction efficiencies of over 75% [71, 72]. Retroviruses and derived vectors integrate into the chromosome of the host cell and allow in this way a stable long-term transgene expression. However, the preference of retroviruses to integrate in transcription-active genomic loci can also result in an interference or destruction of host genes. Studies have revealed a nonrandom preference of MoMuLV integration into fragile chromosomal sites [73]. In addition, retroviral long terminal repeats include enhancer elements which can upregulate cellular genes over long distances of more than 10 kb [74]. Considering the presence of more than 100 proto-oncogenes in the human genome, oncogene dysregulation may occur in about 0.1–1% of all retroviral gene-transfer events [75]. Theoretically the risk of an insertion within 10 kbp of a potential proto-oncogene can be estimated to range between 10^{-2} and 10^{-3}. However, there currently exist 3 known factors for the prevention of such insertion events: first, retrovirus vector insertion is almost uniformly monoallelic; second, some gene insertions may trigger differentiation or apoptosis, or otherwise reduce the survival probability of the affected cell clone; third, a single insertional mutation is, according to current knowledge, not sufficient to develop a malignant phenotype by itself [75].

The advantage of lentiviral transduction is that it allows the expression of the CAR even in nonproliferating cells, including central memory T cells. Since these cells are able to maintain long-term CAR-mediated anticancer immunity, they are of special interest [46]. When a lentiviral system was used it was even possible to transduce stem cell-like memory T cells, a subset of these memory cells, with an antimesothelin CAR [76]. Additionally, the theoretical risk for carcinogenic events is lower compared to retroviruses, because lentiviruses integrate away from host promotors [77]. These theoretical considerations could be confirmed in a pooled analysis of 150 patient years where no mutagenic events were detectable [78]. The lentiviral system, created for clinical use, consists of 4 components: two plasmids coding for packing genes, one coding for the envelope and one coding for the CAR.

In contrast to viral transduction, most plasmid vector-based gene delivery systems do not achieve permanent transfection in primary lymphocytes and many lymphoid cell lines. However, this can be a major advantage for the implementation of novel CAR-based therapies because, if the expression of the CAR is only temporary, the side effects caused by on-target/off-tissue binding of the CAR are also not long lasting. Therefore, transfection systems are recommended for CARs targeting novel antigens with poorly characterized antigen expression in healthy tissue. There exist two possibilities for CAR gene transfer of primary T cells using plasmid vectors: firstly, via electroporation [79, 80] and secondly, via the transposon system Sleeping Beauty [81]. This system is composed of a transposase and a transposon, designed for insertion of specific sequences of DNA into genomes. Transposons translocate DNA from one DNA site to another site in the same or different DNA molecule or genome.

Animal Models – A Difficult Assumption

The antigen-specific cytotoxic effect of CAR gene-modified T cells against tumor cells has been demonstrated in multiple studies in vitro and in vivo (table 1). However, all of the used animal models show major limitations resulting in a difficult assumption of the therapeutic and potentially harmful effects of redirected T cells. The majority of the studies are xenotransplant models allowing the testing of human redirected T cells against human tumor cells. These models do not allow the investigation of the interaction of the host immune system with the transferred redirected T cells because they have to be performed in immune-compromised mice [NSG (NOD.Cg-*Prkdc*^{scid} *Il2rg*^{tm1Wjl}/SzJ), NOD (*scid* – NOD.CB17-*Prkdc*^{scid}/J), NUDE (CAnN.Cg-*Foxn1*^{nu}), SCID (B6.CB17-*Prkdc*^{scid}/SzJ or CBySmn.CB17-*Prkdc*^{scid}/J)]. The observation of in vivo effects was made using so-called protection models. In one study human T cells transduced with a tumor-associated glycoprotein (TAG) 72-specific CAR construct and TAG72+ tumor cells were coinjected subcutaneously into NOD/SCID mice, whereby 3 out of 4 mice survived and did not develop tumors. Furthermore, the intraperitoneal coinjection of these CAR+ T cells inhibited the development of TAG72+ KLE B cell tumors in 4 out of 4 mice [53]. In another model of multiple myeloma using the cell line U266 in NSG mice, 100% of the animals were protected [59]. However, the coinjection of tumor cells and redirection of T cells is not the optimal model to test redirected T cells since new therapy strategies are needed, especially for large inoperable and/or metastasized tumors. A multitude of animal models have been designed to test the ability of redirected T cells to eradicate established tumors (table 1). Alternatively, syngeneic animal tumor models can be used in which the complex interplay between tumor, redirected T cells and the host immune system can be studied [38]. However, the signaling domains and the transmembrane domain have to be adjusted to murine T cells resulting in the necessity to clone new human-adjusted CAR constructs for the translation into clinical trials. Again, the syngeneic models also allow only an assumption of the efficacy of redirected T cells in humans. The ideal model would be in tumor-bearing, humanized mice [82]. To avoid allogeneic recognition, the tumor and the CD34+ stem cells for humanization should derive from the same patient. Furthermore, to assess on-target/on-tissue (therapeutic effects) and on-target/off-tissue (side effects) consequences, the employed CAR should be able to recognize the human and the murine version of the target antigen.

Therapeutic Applications

Production of Redirected T Cells for Clinical Use – A Complex and Expensive Prelude

The procedures for manufacturing redirected T cells and their clinical application includes the following steps (fig. 3): (1) leukapheresis: collection of the patient's T cells from peripheral blood; (2) T cell activation ex vivo: T cells are activated using

Table 1. In vitro and in vivo studies with redirected T cells

Target	CAR	Generation	Disease	Model system	Transduction	Selected references
CD38	scFv anti-CD38-CD3ζ	1	B cell, myeloma	NOD/SCID	Retroviral	[107, 108]
CD19	scFv anti-CD19-CD3ζ	1	B cell	SCID-beige		[109]
CD19	scFv anti-CD19-CD3ζ	1	B cell	in vitro		[110]
CD19	scFv-Fc-CD3ζ	1	Melanoma	Crl:NIH-Lyst^{bg}Foxn1^{nu}Btk^{xid}	Retroviral	[111]
CD20	scFv anti-CD20-FC-CD3ζ	1	B cell	in vitro	Electroporated	[112]
CD22	scFv anti-CD22-CD3ζ	1	B cell	in vitro	Retroviral	[113]
CD30	scFv anti-CD30-CD3ζ	1	Hodgkin's and non-Hodgkin's lymphomas	CB17/SCID	Retroviral	[114, 115]
ROR1	scFv anti-ROR1-CD3ζ	1	B cell	in vitro	Electroporated	[116]
ErbB2	scFv anti-ErbB2-CD3ζ	1	Solid cancer	in vitro	Retroviral	[117]
ErbB3/4	Ligand Erb3/4-CD3ζ	1	Solid cancer	in vitro		[118]
CEA	scFv anti-CEA-CD3ζ	1	Solid cancer	C57BL/6	Retroviral	[28]
CEA	scFv anti-CEA-FcεRI-γ	1	Solid cancer	C57BL/6	Retroviral	[28]
EGP2	scFv anti-EGP2-FcεRI-γ	1	Solid cancer	in vitro	Retroviral	[119]
EGP40	scFv anti-EGP40-FcεRI-γ	1	Colon	in vitro	Retroviral	[120]
TAG72	scFv anti-TAG72-CD28-CD3ζ	1	Gastrointestinal	NOD/SCID	Retroviral	[53]
NKG2D ligands	NKG2Dligand-CD3ζ	1	Several	C57BL/6	Retroviral	[121]
CA9	scFv anti-CA9-FcεRI-γ	1	Renal cell carcinoma	in vitro	Retroviral	[122]
GD2	scFv anti-GD2-CD3ζ	1	Neuroblastoma, Ewing's sarcoma	in vitro	Retroviral	[123]

Table 1. Continued

Target	CAR	Generation	Disease	Model system	Transduction	Selected references
GD3	scFv anti-GD2-CD3ζ	1	Melanoma	in vitro	Retroviral	[21]
HMW-MAA	scFv anti-HMW-MAA-FcεRI-γ	1	Melanoma	Proteoglycan	Retroviral	[124]
CD171	scFv C171-CD3ζ	1	Neuroblastoma	in vitro/clinical trial	Electroporated	[102, 125]
PSCA	scFv anti-PSCA-CD3ζ	1	Prostate	Protein	Retroviral	[126]
Folate receptor-α	scFv anti-FR-FcεRI-γ	1	Ovarian and others	C57BL/6		[127, 128]
CD44v6	scFv anti-CD44-CD3ζ	1	Several	Nude mice	Retroviral	[129]
CD44v7/8	scFv anti-CD44-CD3ζ	1	Cervical	BALB/c	Retroviral	[130]
αvβ6 integrin	Integrin-binding-peptide-CD3ζ	1	Several	in vitro	Electroporated	[131]
NCAM	scFv anti-NCAM-CD3ζ	1	Neuroblastoma	in vitro	Retroviral	[132]
VEGF receptors	VEGF-CD3ζ	1	Several	BALB/c	Retroviral	[133]
Fetal AChR	scFv anti-AChR-CD3ζ	1	Rhabdomyosarcoma	in vitro	Retroviral	[134]
HLA-A2/NY-ESO-1	scFv anti-HLA-A2/NY-ESO-1-CD28-CD3ζ	2	Several	NOG/SCID/γc^{null}	Retroviral	[59,64]
CD19	scFv anti-CD19-CD28-CD3ζ	2	B cell	SCID-Beige		[135]
CD19	scFv anti-CD19-CD3ζ-CD28	2	B cell	in vitro		[136]
CD19	scFv anti-CD19-4-1BB-CD3ζ	2	B cell			
κ light chain	scFv anti-kappa-light-chain-CD28-CD3ζ	2	B cell	SCID-Beige	Retroviral	[137]
CD30	scFv anti-CD30-CD28-CD3ζ	2	Hodgkin's and non-Hodgkin's lymphomas	NOG/SCID/γc^{null}	Retroviral	[138]
CD33	scFv anti-CD30–CD33-CD3ζ	2	Myeloid	NOD/SCID	Retroviral	[139]

Table 1. Continued

Target	CAR	Generation	Disease	Model system	Transduction	Selected references
CD123	scFv anti-CD123-CD28-CD3ζ	2	Myeloid		Electroporated	[140]
ErbB2	scFv anti-ErbB2-CD28-CD3ζ	2	Solid cancer	BALB/c & C57BL/6	Retroviral	[38]
ErbB2	scFv anti-ErbB2-CD28-FcRIγ	2	Solid cancer	CB.17-SCID BEIGE	Retroviral	[141]
Several ErbB dimers	Ligand-panErb- CD28-CD3ζ	2	Solid cancer	SCID Beige	Retroviral	[142]
CEA	scFv anti-CEA-CD28-CD3ζ	2	Solid cancer	C57BL/6	Retroviral	[143]
Mesothelin	scFv anti-Meso-4-1BB-CD3ζ	2	Solid cancer	NSG	Lentiviral	[76]
PSMA	scFv anti-PSMA-CD28-CD3ζ	2	Prostate	in vitro	Retroviral	[144]
B7-H6	NKp30-CD28-CD3ζ	2	Several	C57BL/6	Retroviral	[145]
IL-13 receptor α2	IL13- CD28-CD3ζ	2	Glioblastoma	Nude rats	Retroviral	[146]
MUC16	scFV-anti-MUC16-CD28-CD3ζ	2	Ovarian	SCID Beige	Retroviral	[147]
GD2	scFv anti-GD2-CD28-CD3ζ	2	Neuroblastoma, Ewing's sarcoma	in vitro	Retroviral	[35]
GD2	scFv anti-GD2-CD28-CD3ζ	2	Neuroblastoma, Ewing's sarcoma	NOD/SCID	Retroviral	[148]
Lewis Y	scFv anti-LeY-CD28-CD3ζ	2	Several	NOD/SCID	Retroviral	[149]
HLA-A1 MAGEA1	scFv anti-HLA-A1 MAGE A1-CD28-CD3ζ	2	Melanoma	in vitro	Retroviral	[150]
8H9	scFv8H9-CD28-CD3ζ	2	Solid cancer	SCID-beige	Retroviral	[151]
CD19	scFv anti-CD19-CD8hinge-CD28-4-1BB-CD3ζ	3	B cell	in vitro		[152]

Table 1. Continued

Target	CAR	Generation	Disease	Model system	Transduction	Selected references
CD33	scFv anti-CD33-CD28-OX40-CD3ζ	3	Myeloid	in vitro	Retroviral	[153]
ErbB2	scFv anti- ErbB2(Trastuzumab)-CD28-CD137-CD3z	3	Solid cancer	SCID	Retroviral	[45]
EGFRvIII	scFv anti-EGFRvIII-CD28-41BB-CD3ζ	3	Solid cancer	in vitro	Retroviral	[154]
Mesothelin	scFv anti-Mesothelin-CD28-4-1BB-CD3ζ	3	Solid cancer	NOD/Shi-*scid*/IL-2Rγnull	Lentiviral	[155]
PSMA	scFv anti-PSMA-CD28-4-1BB-CD3ζ	3	Prostate	SCID-beige	Retroviral	[156]
MUC1	scFv-anti-MUC1-CD28-OX40-CD3ζ	3	Breast, ovarian	SCID-beige	Retroviral	[157]
MUC1	scFv-anti-CSPG4-CD28-CD3ζ	2	Melanoma, breast cancer, head and neck squamous cell carcinoma, mesothelioma, glioblastoma, clear cell renal carcinoma, sarcomas	NOD/SCID	Retroviral	[158]

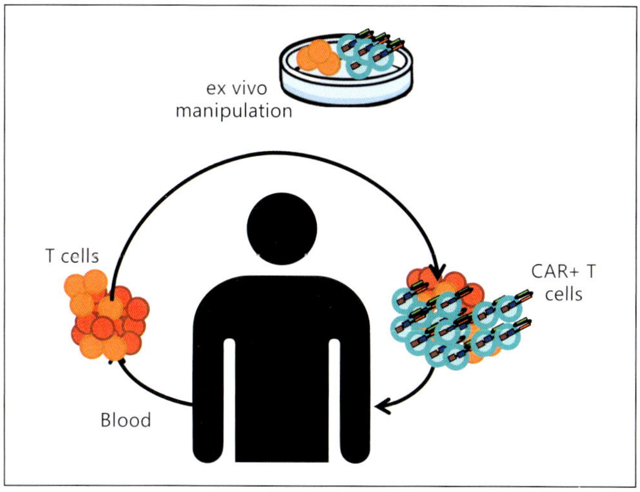

Fig. 3. The process of manufacturing redirected T cells.

antibodies (beads or plate bound); (3) transduction (viral based) or transfection (non-viral) ex vivo: T cells are genetically transduced or transfected with a gene construct encoding the CAR; (4) expansion ex vivo: gene-modified cells undergo further expansion (optional). Depending on the planned numbers of redirected T cells for the adoptive transfer, T cells have to be taken from leukapheresis products or directly from peripheral blood samples. Both collection methods are safe [83, 84]. To enrich T cells from the blood products antibody-coated nanoparticles can be used to separate T cells in general (e.g. CD3) or specific subsets of T cells like memory T cells (CD3 and CD45RO) [85]. Next, the T cells have to be activated to induce proliferation. Again, antibody-coated nanoparticles are used providing stimulation of the CD3 and CD28 receptors. GMP (good manufacturing practice)-grade produced lenti- or retroviruses containing the CAR sequence are used for gene transfer [86]. Afterwards, the transduced T cells can be expanded by providing cytokine cocktails, including IL-2, IL-7 and IL-15. Additionally, the antigen recognized by the CAR can be used to drive CAR-specific expansion of transduced T cells. The antigen can be provided by artificial antigen-presenting cells [87] or anti-idiotypic antibodies [88]. The final formulation as genetically modified transplant product can be frozen and stored until clinical use. Since all the described steps have to be performed under GMP and clean room conditions, the process of commercialization and academic distribution are very cost-, labor- and time-intensive, resulting in the slow spread of this cellular therapy among medical centers.

Clinical Trials with Redirected T Cells – The Proof of Concept Has Just Been Provided
Regarding clinical development, the focus lies especially on B cell-derived malignancies because the targets CD19 and CD20 are well characterized and on-target/off-tissue side effects (mainly physiological B cell) have been well described in anti-CD20 antibody treatment. Thus, this setting offers many possibilities for testing more innovative approaches. One way is through the improvement of cell sources. In this

Redirected T Cells Targeting B Cell Neoplasia – The Prime Example

As mentioned before, the most developed T cell therapy with redirected T cells is the treatment against CD19. All B cells (except hematopoietic stem cells and plasma cells) express CD19 at all stages of development, allowing the treatment of a wide range of B cell neoplasia. However, physiological B cells will also be eradicated, leading to the same side effects of anti-CD20 antibodies that have been known for many years in the clinic.

In the first clinical trial with redirected T cells in B cell lymphoma, an anti-CD20-CD3ζ CAR was used [90]. Transduction was performed through electroporation and patients received up to 4.4×10^9 cells intravenously over 5 days. To achieve this number, bulk-transduced cells were administered. The last cohort of 4 patients received a combination therapy consisting of lymphodepleting chemotherapy and IL-2 to stimulate T cell survival. Redirected T cells were detectable up to 9 weeks after transfer, and no grade 3 or 4 toxicities were observed.

Preclinical data clearly indicated the need for costimulatory signals in redirected T cells to allow T cell survival and to improve function. Therefore, second-generation CARs were tested basically following two concepts: CARs using an 'early' costimulatory signal (CD28) and CARs with 'late' costimulatory signals (4-1BB). Savoldo et al. [91] performed an important clinical comparison of first- and second-generation CARs (anti-CD19) to test whether CD28 as an early costimulatory signal improves the survival of redirected T cells. In this trial autologous redirected T cells expressing first- or second-generation CARs were simultaneously transferred, without the addition of lympho-depleting chemotherapy. The CD28 costimulatory signal induced proliferation in vivo, indicating superior survival through an 'early' costimulatory signal. Furthermore, the group at Memorial Sloan Kettering Cancer Center in New York tested a CD19-CD28-CD3ζ CAR [92]. In this study the first cohort received redirected T cells without lympho-depleting chemotherapy. The redirection of T cells was performed by retroviral transfer. In the absence of any signs of toxicity in the first 3 patients, the fourth patient received lympho-depleting chemotherapy (cyclophosphamide 1.5 g/m^2) followed by adoptive transfer of redirected T cells. The patient died of multiorgan failure after a cytokine storm [93]. This suspected unexpected serious adverse reaction (SUSAR) was finally interpreted as septic shock caused by an underlying infection. Five further patients showed a good tolerance of lympho-depletion and adoptive transfer of redirected T cells expressing CD19-CD28-CD3ζ CARs. However, this event led to an intensive discussion between investigators and regulatory authorities about the toxicity of redirected T cells and the introduction of appropriate safety measures for clinical trials in this field. Additionally, other multiple side effects, declared as not severe, were observed, including fever, rigors and hypotension. In 1 patient persisting B cell aplasia was observed and was considered to be a nonsevere expected on-target/off-tissue effect. Regarding efficacy, no complete response could be induced in this study; 1 patient with chronic lymphatic leukemia (CLL) showed a partial remission. At the US National Institute of Health (NIH) in Bethesda, a clinical trial with CD19-CD28-CD3ζ CARs combined with lympho-depleting chemotherapy and IL-2 as a cytokine support was performed. Six out of 8 patients showed a clinical response. The side effects were fever, hypotension, renal failure and fatigue [94].

Alternatively, 'late' costimulation was induced by a CD19-4-1BB-CD3ζ. In this case CARs were transfected by lentiviral transfer and tested in patients with chronic lymphatic leukemia receiving lympho-depleting chemotherapy [47]. Even though some patients received only 1.42×10^7 CD3+ redirected T cells, they showed clinical responses, and expansion of redirected T cells and memory formation was seen [46]. Patients developed a tumor lysis syndrome, which was interpreted on the one hand as a sign of clinical response, and on the other hand was viewed as a severe side effect. Again, B cell aplasia was induced. At least for treating B cell-derived malignancies this CAR design (CD19-4-1BB-CD3ζ CAR) was a major breakthrough resulting in increased interest from academia and industry to further develop the adoptive transfer of CAR redirected T cells. The following anti-CD19 CAR was also tested in patients with acute lymphatic leukemia and showed impressive results. Here, two constructs from two different research groups – the CD19-4-1BB-CD3ζ [95] and the CD19-CD28-CD3ζ [96] CAR – induced complete remissions, with the CD19-4-1BB-CD3ζ CAR seeming to be responsible for more long-term remissions. Also, third-generation CARs were explored in redirected T cells (CD19-CD28-4-1BB-CD3ζ) in combination with lympho-depleting chemotherapy and IL-2 [44]. The CARs were transfected by electroporation. No severe side effects were observed. Only 1 patient experienced fever, hypotension and respiratory distress.

Clinical Studies with CAR-Modified T Lymphocytes Targeting Solid Cancers – The Challenge

In solid cancers the translation of redirected T cells from bench to bedside represents a special challenge. The major difficulty consists in finding the right target antigen allowing on-tissue/on-target without off-tissue/on-target effects. Further homing to the tumor of the redirected T cells is needed. In 2006 a sentinel phase I/II trial was reported in patients with metastatic renal cell cancer. Redirected T cells were generated by transfection with a CAR consisting of an scFv derived from the carbonic anhydrase IX (CAIX)-specific murine monoclonal antibody G250, the CD4 transmembrane domains and the FcεRI γ signal chain [97]. All patients were treated with the transduced T cells in an in-patient dose-escalation scheme of intravenous doses of 2×10^7 cells on day 1, 2×10^8 cells on day 2, 2×10^9 cells on days 3–5 (cycle 1), and 2×10^9 cells on days 17–19 (cycle 2) in combination with IL-2 (5×10^5 U/m^2 twice daily administered subcutaneously on days 1–10 and days 17–26) [29]. Redirected T cells were detected in the circulation up to 32 days by flow cytometry and up to 53 days by polymerase chain reaction (PCR) [98]. Initially, infusions of these gene-modified T cells were well tolerated, but after 4–5 infusions liver enzyme disturbances developed, reaching National Cancer Institute (NCI) Common Toxicity Criteria grades 2–4. These were probably caused by an on-target/off-tissue anti-CAIX T cell response against the bile duct epithelial cells. Alternatively, toxicity could also be unspecific due to the biodistribution of activated T cells, which can be found in the liver after passaging the lung [99]. Additionally, all 3 patients developed low levels of anti-scFv(G250)-γ antibodies between 37 and 100 days after the start of T cell therapy, which were directed against the G250 idiotype. Furthermore, antibodies against the plasmid used for retro-viral transfer were found [100]. To prevent liver toxicity in subsequent patients, 3 days before the first infusion of anti-CAIX redirected T cells, 5 mg of chimeric G250 antibody was administered. Liver toxicities were thereby avoided and enhanced peripheral T cell persistence was observed [101].

Since virus-specific T cells show the advantage of effector and memory capacity, Epstein-Barr virus (EBV)-specific T cells were transduced with first-generation anti-GD2-CD3ζ CAR. To test whether EBV-specific redirected T cells were superior to polyclonal redirected T cells, both populations were intraindividually analyzed in a phase I study including 19 glioblastoma patients [102]. Virus-specific T cells expanded more vigorously; however, both populations of redirected T cells had faded at the latest by 6 weeks after transfer. Interestingly, EBV-specific T cells could be found by PCR up to 96 weeks, whereas polyclonal redirected T cells could be found up to 192 weeks after transfer. Complete remissions were seen in 3 patients. In solid cancers showing positivity for ErbB2, a third-generation CAR has already been tested [anti-ErbB2 (trastuzumab)-CD28-4-1BB-CD3ζ CAR]. One patient with ErbB2-positive colorectal cancer was treated with 1×10^{10} cells and lympho-depleting chemotherapy. The patient experienced immediate severe respiratory distress and died 5 days later. The fatal side effect was explained by an on-target/off-tissue effect caused by low-level expression of ErbB2 on the lung tissue [103]. ErbB2 as a target was further evaluated by using a second-generation CAR (anti-ErbB2-CD28-CD3ζ). A clinical trial (NCT00902044) was performed with initial low doses of CAR T cells (1×10^4 CAR+ T cells/m^2) and escalated to 1×10^8 CAR+ T cells/m^2. There was no dose-limiting toxicity observed and the median overall survival was 10.3 months [104]. The two major differences between the two approaches were, firstly, the CAR construct (third- versus second-generation CAR) and, secondly, the antibody used to recognize ErbB2 (trastuzumab versus FRP5). This resulted in one approach with and one without toxicity. In order to improve safety neither lympho-depletion nor cytokine support was administered for the FRP5 CAR.

In solid tumors a strictly conservative strategy has first to be followed before allowing the development of truly therapeutic approaches. In line with that strategy anti-mesothelin redirected T cells were tested in patients with mesothelioma and pancreatic cancer (NCT01355965). In this study safety was improved by transient expression of the CAR after electroporation of the gene construct encoding the chimeric receptor [105].

Conclusions – Redirected T Cells Are Cellular Drivers in Modern Medicine

It seems that now, after 20 years of extensive preclinical research and early clinical experience, the time point has arrived where genetically modified T cells have successfully entered the scene of modern oncology. The knowledge, the techniques and the regulatory structure have been acquired to push therapy with redirected T cells to become part of standard medical treatment. At this moment in time the next critical step is the commercialization of redirected T cells as a medical product. Most likely, the first redirected T cell product will target B cell malignancies. However, from a health and commercial point of view this is only a small patient population and a much greater need lies in the development of redirected T cells against solid cancers, like prostate, breast and lung cancer. Due to the difficulty in finding safe targets, this will require a greater effort of intense research in the next decade. From a conceptual point of view there is a very attractive target for redirected T cells in solid cancer, which could be the ultimate safe target: neoantigens. Since the first description by Schumacher and Schreiber [106], increasing evidence has emerged of their expression in malignant tumors, lately clinically reflected by the observation that checkpoint inhibition results in better clinical responses in highly mutated melanomas, presumably expressing more neoantigens. Therefore – at least in part – the development of redirected T cells will be linked closely to the technical development allowing the production of neoantigen-specific CARs for an individual patient. CAR redirected T cells would thereby become an absolutely personalized cellular therapeutic.

References

1 Scott AM, Wolchok JD, Old LJ: Antibody therapy of cancer. Nat Rev Cancer 2012;12:278–287.

2 Burnet FM: The concept of immunological surveillance. Prog Exp Tumor Res 1970;13:1–27.

3 Smyth MJ, Godfrey DI, Trapani JA: A fresh look at tumor immunosurveillance and immunotherapy. Nat Immunol 2001;2:293–299.

4 Smyth MJ, Dunn GP, Schreiber RD: Cancer immunosurveillance and immunoediting: the roles of immunity in suppressing tumor development and shaping tumor immunogenicity. Adv Immunol 2006;90:1–50.

5 Perez P, Hoffman RW, Shaw S, Bluestone JA, et al: Specific targeting of cytotoxic T cells by anti-T3 linked to anti-target cell antibody. Nature 1985;316:354–356.

6 Canevari S, Stoter G, Arienti F, Bolis G, et al: Regression of advanced ovarian-carcinoma by intraperitoneal treatment with autologous T-lymphocytes retargeted by a bispecific monoclonal antibody. J Natl Cancer Inst 1995;87:1463–1469.

7 Jain RK: Physiological barriers to delivery of monoclonal-antibodies and other macromolecules in tumors. Cancer Res 1990;50:S814–S819.

8 Blankvoorthuis CJAC, Braakman E, Ronteltap CPM, Tilly BC, et al: Clustered CD3 TCR complexes do not transduce activation signals after bispecific monoclonal antibody-triggered lysis by cytotoxic T-lymphocytes via CD3. J Immunol 1993;151:2904–2914.

9 Kipriyanov SM, Cochlovius B, Schafer HJ, Moldenhauer G, et al: Synergistic antitumor effect of bispecific CD19 × CD3 and CD19 × CD16 diabodies in a preclinical model of non-Hodgkin's lymphoma. J Immunol 2002;169:137–144.

10 Bargou R, Leo E, Zugmaier G, Klinger M, et al: Tumor regression in cancer patients by very low doses of a T cell-engaging antibody. Science 2008;321:974–977.

11 Topp MS, Gokbuget N, Zugmaier G, Degenhard E, et al: Long-term follow-up of hematologic relapse-free survival in a phase 2 study of blinatumomab in patients with MRD in B-lineage ALL. Blood 2012;120:5185–5187.

12 Topp MS, Kufer P, Gokbuget N, Goebeler M, et al: Targeted therapy with the T-cell-engaging antibody blinatumomab of chemotherapy-refractory minimal residual disease in B-lineage acute lymphoblastic leukemia patients results in high response rate and prolonged leukemia-free survival. J Clin Oncol 2011;29:2493–2498.

13 Eshhar Z, Bach N, Fitzer-Attas CJ, Gross G, et al: The T-body approach: potential for cancer immunotherapy. Springer Semin Immunopathol 1996;18:199–209.

14 Restifo NP, Dudley ME, Rosenberg SA: Adoptive immunotherapy for cancer: harnessing the T cell response. Nat Rev Immunol 2012;12:269–281.

15 Hombach AA, Holzinger A, Abken H: The weal and woe of costimulation in the adoptive therapy of cancer with chimeric antigen receptor (CAR)-redirected T cells. Curr Mol Med 2013;13:1079–1088.
16 Kuwana Y, Asakura Y, Utsunomiya N, Nakanishi M, et al: Expression of chimeric receptor composed of immunoglobulin-derived V regions and T-Cell receptor-derived C regions. Biochem Biophys Res Commun 1987;149:960–968.
17 Goverman J, Gomez SM, Segesman KD, Hunkapiller T, et al: Chimeric immunoglobulin-T cell receptor proteins form functional receptors: implications for T cell receptor complex formation and activation. Cell 1990;60:929–939.
18 Roberts MR, Qin L, Zhang D, Smith DH, et al: Targeting of human immunodeficiency virus-infected cells by CD8+ T lymphocytes armed with universal T-cell receptors. Blood 1994;84:2878–2889.
19 Scholler J, Brady TL, Binder-Scholl G, Hwang WT, et al: Decade-long safety and function of retroviral-modified chimeric antigen receptor T cells. Sci Transl Med 2012;4:132ra53.
20 Nolan KF, Yun CO, Akamatsu Y, Murphy JC, et al: Bypassing immunization: optimized design of 'designer T cells' against carcinoembryonic antigen (CEA)-expressing tumors, and lack of suppression by soluble CEA. Clin Cancer Res 1999;5:3928–3941.
21 Yun CO, Nolan KF, Beecham EJ, Reisfeld RA, et al: Targeting of T lymphocytes to melanoma cells through chimeric anti-GD3 immunoglobulin T-cell receptors. Neoplasia 2000;2:449–459.
22 Bird RE, Hardman KD, Jacobson JW, Johnson S, et al: Single-chain antigen-binding proteins. Science 1988;242:423–426.
23 Asano R, Takemura S, Tsumoto K, Sakurai N, et al: Functional construction of the anti-mucin core protein (MUC1) antibody MUSE11 variable regions in a bacterial expression system. J Biochem (Tokyo) 2000;127:673–679.
24 Gilham DE, Debets R, Pule M, Hawkins RE, et al: CAR-T cells and solid tumors: tuning T cells to challenge an inveterate foe. Trends Mol Med 2012;18: 377–384.
25 Eshhar Z, Fitzer-Attas CJ: Tyrosine kinase chimeras for antigen-selective T-body therapy. Adv Drug Deliv Rev 1998;31:171–182.
26 Zenner G, Vorherr T, Mustelin T, Burn P: Differential and multiple binding of signal transducing molecules to the ITAMs of the TCR-ζ chain. J Cell Biochem 1996;63:94–103.
27 Shores E, Flamand V, Tran T, Grinberg A, et al: FcεRI γ can support T cell development and function in mice lacking endogenous TCR ζ-chain. J Immunol 1997;159:222–230.
28 Haynes NM, Snook MB, Trapani JA, Cerruti L, et al: Redirecting mouse CTL against colon carcinoma: superior signaling efficacy of single-chain variable domain chimeras containing TCR-ζ vs FcεRI-γ. J Immunol 2001;166:182–187.
29 Lamers CH, Sleijfer S, Vulto AG, Kruit WH, et al: Treatment of metastatic renal cell carcinoma with autologous T-lymphocytes genetically retargeted against carbonic anhydrase IX: first clinical experience. J Clin Oncol 2006;24:e20–e22.
30 Kershaw MH, Westwood JA, Parker LL, Wang G, et al: A phase I study on adoptive immunotherapy using gene-modified T cells for ovarian cancer. Clin Cancer Res 2006;12:6106–6115.
31 Finney HM, Lawson AD, Bebbington CR, Weir AN: Chimeric receptors providing both primary and costimulatory signaling in T cells from a single gene product. J Immunol 1998;161:2791–2797.
32 Finney HM, Akbar AN, Lawson AD: Activation of resting human primary T cells with chimeric receptors: costimulation from CD28, inducible costimulator, CD134, and CD137 in series with signals from the TCRζ chain. J Immunol 2004;172:104–113.
33 Friedmann-Morvinski D, Bendavid A, Waks T, Schindler D, et al: Redirected primary T cells harboring a chimeric receptor require costimulation for their antigen-specific activation. Blood 2005;105: 3087–3093.
34 Hombach A, Sent D, Schneider C, Heuser C, et al: T-cell activation by recombinant receptors: CD28 costimulation is required for interleukin 2 secretion and receptor-mediated T-cell proliferation but does not affect receptor-mediated target cell lysis. Cancer Res 2001;61:1976–1982.
35 Krause A, Guo HF, Latouche JB, Tan C, et al: Antigen-dependent CD28 signaling selectively enhances survival and proliferation in genetically modified activated human primary T lymphocytes. J Exp Med 1998;188:619–626.
36 Alvarez-Vallina L, Hawkins RE: Antigen-specific targeting of CD28-mediated T cell co-stimulation using chimeric single-chain antibody variable fragment-CD28 receptors. Eur J Immunol 1996;26: 2304–2309.
37 Haynes NM, Trapani JA, Teng MW, Jackson JT, et al: Rejection of syngeneic colon carcinoma by CTLs expressing single-chain antibody receptors codelivering CD28 costimulation. J Immunol 2002;169: 5780–5786.
38 Haynes NM, Trapani JA, Teng MW, Jackson JT, et al: Single-chain antigen recognition receptors that costimulate potent rejection of established experimental tumors. Blood 2002;100:3155–3163.

39 Ren-Heidenreich L, Mordini R, Hayman GT, Siebenlist R, et al: Comparison of the TCR ζ-chain with the FcR γ-chain in chimeric TCR constructs for T cell activation and apoptosis. Cancer Immunol Immunother 2002;51:417–423.
40 Kofler DM, Chmielewski M, Rappl G, Hombach A, et al: CD28 costimulation impairs the efficacy of a redirected T-cell antitumor attack in the presence of regulatory T cells which can be overcome by preventing Lck activation. Mol Ther 2011;19:760–767.
41 Gramaglia I, Weinberg AD, Lemon M, Croft M: Ox-40 ligand: a potent costimulatory molecule for sustaining primary CD4 T cell responses. J Immunol 1998;161:6510–6517.
42 Rogers PR, Song J, Gramaglia I, Killeen N, et al: OX40 promotes Bcl-xL and Bcl-2 expression and is essential for long-term survival of CD4 T cells. Immunity 2001;15:445–455.
43 Pule MA, Straathof KC, Dotti G, Heslop HE, et al: A chimeric T cell antigen receptor that augments cytokine release and supports clonal expansion of primary human T cells. Mol Ther 2005;12:933–941.
44 Till BG, Jensen MC, Wang J, Qian X, et al: CD20-specific adoptive immunotherapy for lymphoma using a chimeric antigen receptor with both CD28 and 4-1BB domains: pilot clinical trial results. Blood 2012;119:3940–3950.
45 Zhao Y, Wang QJ, Yang S, Kochenderfer JN, et al: A herceptin-based chimeric antigen receptor with modified signaling domains leads to enhanced survival of transduced T lymphocytes and antitumor activity. J Immunol 2009;183:5563–5574.
46 Kalos M, Levine BL, Porter DL, Katz S, et al: T cells with chimeric antigen receptors have potent antitumor effects and can establish memory in patients with advanced leukemia. Sci Transl Med 2011;3:95ra73.
47 Porter DL, Levine BL, Kalos M, Bagg A, et al: Chimeric antigen receptor-modified T cells in chronic lymphoid leukemia. N Engl J Med 2011;365:725–733.
48 Kloss CC, Condomines M, Cartellieri M, Bachmann M, et al: Combinatorial antigen recognition with balanced signaling promotes selective tumor eradication by engineered T cells. Nat Biotechnol 2013;31:71–75.
49 Hanada K, Restifo NP: Double or nothing on cancer immunotherapy. Nat Biotechnol 2013;31:33–34.
50 Willemsen RA, Debets R, Chames P, Bolhuis RL: Genetic engineering of T cell specificity for immunotherapy of cancer. Hum Immunol 2003;64:56–68.
51 Romeo C, Amiot M, Seed B: Sequence requirements for induction of cytolysis by the T cell $^{antigen}/_{Fc}$ receptor ζ chain. Cell 1992;68:889–897.
52 Fitzer-Attas CJ, Schindler DG, Waks T, Eshhar Z: Harnessing Syk family tyrosine kinases as signaling domains for chimeric single chain of the variable domain receptors: optimal design for T cell activation. J Immunol 1998;160:145–154.
53 McGuinness RP, Ge Y, Patel SD, Kashmiri SV, et al: Anti-tumor activity of human T cells expressing the CC49-ζ chimeric immune receptor. Hum Gene Ther 1999;10:165–173.
54 Kolanus W, Romeo C, Seed B: T cell activation by clustered tyrosine kinases. Cell 1993;74:171–183.
55 Eshhar Z, Waks T, Gross G, Schindler DG: Specific activation and targeting of cytotoxic lymphocytes through chimeric single chains consisting of antibody-binding domains and the γ or ζ subunits of the immunoglobulin and T-cell receptors. Proc Natl Acad Sci U S A 1993;90:720–724.
56 Patel SD, Moskalenko M, Smith D, Maske B, et al: Impact of chimeric immune receptor extracellular protein domains on T cell function. Gene Ther 1999;6:412–419.
57 Moritz D, Groner B: A spacer region between the single chain antibody- and the CD3 ζ-chain domain of chimeric T cell receptor components is required for efficient ligand binding and signaling activity. Gene Ther 1995;2:539–546.
58 Schirrmann T, Pecher G: Specific targeting of $CD33^+$ leukemia cells by a natural killer cell line modified with a chimeric receptor. Leuk Res 2005;29:301–306.
59 Schuberth PC, Jakka G, Jensen SM, Wadle A, et al: Effector memory and central memory NY-ESO-1-specific re-directed T cells for treatment of multiple myeloma. Gene Ther 2013;20:386–395.
60 Hombach A, Sircar R, Heuser C, Tillmann T, et al: Chimeric anti-TAG72 receptors with immunoglobulin constant Fc domains and γ or ζ signalling chains. Int J Mol Med 1998;2:99–103.
61 Hombach A, Hombach AA, Abken H: Adoptive immunotherapy with genetically engineered T cells: modification of the IgG1 Fc ‘spacer’ domain in the extracellular moiety of chimeric antigen receptors avoids ‘off-target’ activation and unintended initiation of an innate immune response. Gene Ther 2010;17:1206–1213.
62 Ren Heidenreich L, Hayman GT, Trevor KT: Specific targeting of EGP-2+ tumor cells by primary lymphocytes modified with chimeric T cell receptors. Hum Gene Ther 2000;11:9–19.
63 Eshhar Z, Waks T, Bendavid A, Schindler DG: Functional expression of chimeric receptor genes in human T cells. J Immunol Methods 2001;248:67–76.
64 Stewart-Jones G, Wadle A, Hombach A, Shenderov E, et al: Rational development of high-affinity T-cell receptor-like antibodies. Proc Natl Acad Sci U S A 2009;106:5784–5788.

65 Finer MH, Dull TJ, Qin L, Farson D, et al: kat: a high-efficiency retroviral transduction system for primary human T lymphocytes. Blood 1994;83:43–50.
66 Weijtens ME, Willemsen RA, Hart EH, Bolhuis RL: A retroviral vector system 'STITCH' in combination with an optimized single chain antibody chimeric receptor gene structure allows efficient gene transduction and expression in human T lymphocytes. Gene Ther 1998;5:1195–1203.
67 Engels B, Cam H, Schuler T, Indraccolo S, et al: Retroviral vectors for high-level transgene expression in T lymphocytes. Hum Gene Ther 2003;14:1155–1168.
68 Gladow M, Becker C, Blankenstein T, Uckert W: MLV-10A1 retrovirus pseudotype efficiently transduces primary human CD4+ T lymphocytes. J Gene Med 2000;2:409–415.
69 Stitz J, Buchholz CJ, Engelstadter M, Uckert W, et al: Lentiviral vectors pseudotyped with envelope glycoproteins derived from gibbon ape leukemia virus and murine leukemia virus 10A1. Virology 2000;273:16–20.
70 Uckert W, Becker C, Gladow M, Klein D, et al: Efficient gene transfer into primary human CD8+ T lymphocytes by MuLV-10A1 retrovirus pseudotype. Hum Gene Ther 2000;11:1005–1014.
71 Altenschmidt U, Klundt E, Groner B: Adoptive transfer of in vitro-targeted, activated T lymphocytes results in total tumor regression. J Immunol 1997; 159:5509–5515.
72 Altenschmidt U, Kahl R, Moritz D, Schnierle BS, et al: Cytolysis of tumor cells expressing the Neu/erbB-2, erbB-3, and erbB-4 receptors by genetically targeted naive T lymphocytes. Clin Cancer Res 1996;2: 1001–1008.
73 Bester AC, Schwartz M, Schmidt M, Garrigue A, et al: Fragile sites are preferential targets for integrations of MLV vectors in gene therapy. Gene Ther 2006;13:1057–1059.
74 Williams DA, Baum C: Gene therapy – new challenges ahead. Science 2003;302:400–401.
75 Baum C, Dullmann J, Li Z, Fehse B, et al: Side effects of retroviral gene transfer into hematopoietic stem cells. Blood 2003;101:2099–2114.
76 Gattinoni L, Lugli E, Ji Y, Pos Z, et al: A human memory T cell subset with stem cell-like properties. Nat Med 2011;17:1290–1297.
77 Durand S, Cimarelli A: The inside out of lentiviral vectors. Viruses 2011;3:132–159.
78 Scholler J, Brady TL, Binder-Scholl G, Hwang WT, et al: Decade-long safety and function of retroviral-modified chimeric antigen receptor T cells. Science Transl Med 2012;4:132ra53.
79 Jensen MC, Clarke P, Tan G, Wright C, et al: Human T lymphocyte genetic modification with naked DNA. Mol Ther 2000;1:49–55.
80 Jensen MC, Cooper LJ, Wu AM, Forman SJ, et al: Engineered CD20-specific primary human cytotoxic T lymphocytes for targeting B-cell malignancy. Cytotherapy 2003;5:131–138.
81 Maiti SN, Huls H, Singh H, Dawson M, et al: Sleeping Beauty system to redirect T-cell specificity for human applications. J Immunother 2013;36:112–123.
82 Rongvaux A, Willinger T, Martinek J, Strowig T, et al: Development and function of human innate immune cells in a humanized mouse model. Nat Biotechnol 2014;32:364–372.
83 Yuan S, Ziman A, Smeltzer B, Lu Q, et al: Moderate and severe adverse events associated with apheresis donations: incidences and risk factors. Transfusion 2010;50:478–486.
84 Winters JL: Complications of donor apheresis. J Clin Apher 2006;21:132–141.
85 Casati A, Varghaei-Nahvi A, Feldman SA, Assenmacher M, et al: Clinical-scale selection and viral transduction of human naive and central memory CD8+ T cells for adoptive cell therapy of cancer patients. Cancer Immunol Immunother 2013;62:1563–1573.
86 Kaiser AD, Assenmacher M, Schroder B, Meyer M, et al: Towards a commercial process for the manufacture of genetically modified T cells for therapy. Cancer Gene Ther 2015;22:72–78.
87 Bajgain P, Mucharla R, Anurathapan U, Lapteva N, et al: A novel approach to manufacture CAR-T cells for clinical applications. Biol Blood Marrow Transplant 2013;19:S194–S195.
88 Jakka G, Schuberth PC, Thiel M, Held G, et al: Antigen-specific in vitro expansion of functional redirected NY-ESO-1-specific human CD8+ T-cells in a cell-free system. Anticancer Res 2013;33:4189–4201.
89 Ertl HC, Zaia J, Rosenberg SA, June CH, et al: Considerations for the clinical application of chimeric antigen receptor T cells: observations from a recombinant DNA Advisory Committee Symposium held June 15, 2010. Cancer Res 2011;71:3175–3181.
90 Till BG, Jensen MC, Wang J, Chen EY, et al: Adoptive immunotherapy for indolent non-Hodgkin lymphoma and mantle cell lymphoma using genetically modified autologous CD20-specific T cells. Blood 2008;112:2261–2271.
91 Savoldo B, Ramos CA, Liu E, Mims MP, et al: CD28 costimulation improves expansion and persistence of chimeric antigen receptor-modified T cells in lymphoma patients. J Clin Invest 2011;121:1822–1826.
92 Brentjens RJ, Riviere I, Park JH, Davila ML, et al: Safety and persistence of adoptively transferred autologous CD19-targeted T cells in patients with relapsed or chemotherapy refractory B-cell leukemias. Blood 2011;118:4817–4828.

93 Brentjens R, Yeh R, Bernal Y, Riviere I, et al: Treatment of chronic lymphocytic leukemia with genetically targeted autologous T cells: case report of an unforeseen adverse event in a phase I clinical trial. Mol Ther 2010;18:666–668.
94 Kochenderfer JN, Dudley ME, Feldman SA, Wilson WH, et al: B-cell depletion and remissions of malignancy along with cytokine-associated toxicity in a clinical trial of anti-CD19 chimeric-antigen-receptor-transduced T cells. Blood 2012;119:2709–2720.
95 Maude SL, Frey N, Shaw PA, Aplenc R, et al: Chimeric antigen receptor T cells for sustained remissions in leukemia. N Engl J Med 2014;371:1507–1517.
96 Lee DW, Kochenderfer JN, Stetler-Stevenson M, Cui YK, et al: T cells expressing CD19 chimeric antigen receptors for acute lymphoblastic leukaemia in children and young adults: a phase 1 dose-escalation trial. Lancet 2015;385:517–528.
97 Bolhuis RL, Willemsen RA, Lamers CH, Stam K, et al: Preparation for a phase I/II study using autologous gene modified T lymphocytes for treatment of metastatic renal cancer patients. Adv Exp Med Biol 1998;451:547–555.
98 Lamers CH, Gratama JW, Pouw NM, Langeveld SC, et al: Parallel detection of transduced T lymphocytes after immunogene therapy of renal cell cancer by flow cytometry and real-time polymerase chain reaction: implications for loss of transgene expression. Hum Gene Ther 2005;16:1452–1462.
99 Parente-Pereira AC, Burnet J, Ellison D, Foster J, et al: Trafficking of CAR-engineered human T cells following regional or systemic adoptive transfer in SCID beige mice. J Clin Immunol 2011;31:710–718.
100 Lamers CH, Willemsen R, van Elzakker P, van Steenbergen-Langeveld S, et al: Immune responses to transgene and retroviral vector in patients treated with ex vivo-engineered T cells. Blood 2011;117:72–82.
101 Lamers CH, Sleijfer S, van Steenbergen S, van Elzakker P, et al: Treatment of metastatic renal cell carcinoma with CAIX CAR-engineered T cells: clinical evaluation and management of on-target toxicity. Mol Ther 2013;21:904–912.
102 Park JR, Digiusto DL, Slovak M, Wright C, et al: Adoptive transfer of chimeric antigen receptor re-directed cytolytic T lymphocyte clones in patients with neuroblastoma. Mol Ther 2007;15:825–833.
103 Morgan RA, Yang JC, Kitano M, Dudley ME, et al: Case report of a serious adverse event following the administration of T cells transduced with a chimeric antigen receptor recognizing ERBB2. Mol Ther 2010;18:843–851.
104 Ahmed N, Brawley VS, Hegde M, Robertson C, et al: Human epidermal growth factor receptor 2 (HER2)-specific chimeric antigen receptor-modified T cells for the immunotherapy of HER2-positive sarcoma. J Clin Oncol 2015;33:1688–1696.
105 Beatty GL, Haas AR, Maus MV, Torigian DA, et al: Mesothelin-specific chimeric antigen receptor mRNA-engineered T cells induce anti-tumor activity in solid malignancies. Cancer Immunol Res 2014;2:112–120.
106 Schumacher TN, Schreiber RD: Neoantigens in cancer immunotherapy. Science 2015;348:69–74.
107 Mihara K, Yanagihara K, Takigahira M, Imai C, et al: Activated T-cell-mediated immunotherapy with a chimeric receptor against CD38 in B-cell non-Hodgkin lymphoma. J Immunother 2009;32:737–743.
108 Mihara K, Bhattacharyya J, Kitanaka A, Yanagihara K, et al: T-cell immunotherapy with a chimeric receptor against CD38 is effective in eliminating myeloma cells. Leukemia 2012;26:365–367.
109 Brentjens RJ, Latouche JB, Santos E, Marti F, et al: Eradication of systemic B-cell tumors by genetically targeted human T lymphocytes co-stimulated by CD80 and interleukin-15. Nat Med 2003;9:279–286.
110 Cooper LJN, Topp MS, Serrano LM, Gonzalez S, et al: T-cell clones can be rendered specific for CD19: toward the selective augmentation of the graft-versus-B-lineage leukemia effect. Blood 2003;101:1637–1644.
111 Schmidt P, Kopecky C, Hombach A, Zigrino P, et al: Eradication of melanomas by targeted elimination of a minor subset of tumor cells. Proc Natl Acad Sci U S A 2011;108:2474–2479.
112 Wang J, Press OW, Lindgren CG, Greenberg P, et al: Cellular immunotherapy for follicular lymphoma using genetically modified CD20-specific CD8+ cytotoxic T lymphocytes. Mol Ther 2004;9:577–586.
113 James ST, Greenberg PD, Jensen MC, Lin YK, et al: Antigen sensitivity of CD22-specific chimeric TCR is modulated by target epitope distance from the cell membrane. J Immunol 2008;180:7028–7038.
114 Hombach A, Heuser C, Sircar R, Tillmann T, et al: An anti-CD30 chimeric receptor that mediates CD3-ζ-independent T-cell activation against Hodgkin's lymphoma cells in the presence of soluble CD30. Cancer Res 1998;58:1116–1119.
115 Savoldo B, Rooney CM, Di Stasi A, Abken H, et al: Epstein Barr virus-specific cytotoxic T lymphocytes expressing the anti-CD30ζ artificial chimeric T-cell receptor for immunotherapy of Hodgkin disease. Blood 2007;110:2620–2630.

116 Hudecek M, Schmitt TM, Baskar S, Lupo-Stanghellini MT, et al: The B-cell tumor-associated antigen ROR1 can be targeted with T cells modified to express a ROR1-specific chimeric antigen receptor. Blood 2010;116:4532–4541.
117 Chmielewski M, Hombach A, Heuser C, Adams GP, et al: T cell activation by antibody-like immunoreceptors: increase in affinity of the single-chain fragment domain above threshold does not increase T cell activation against antigen-positive target cells but decreases selectivity. J Immunol 2004;173:7647–7653.
118 Muniappan A, Banapour B, Lebkowski J, Talib S: Ligand-mediated cytolysis of tumor cells: use of heregulin- ζ chimeras to redirect cytotoxic T lymphocytes. Cancer Gene Ther 2000;7:128–134.
119 Ren-Heidenreich L, Hayman GT, Trevor KT: Specific targeting of EGP-2+ tumor cells by primary lymphocytes modified with chimeric T cell receptors. Hum Gene Ther 2000;11:9–19.
120 Daly T, Royal RE, Kershaw MH, Treisman J, et al: Recognition of human colon cancer by T cells transduced with a chimeric receptor gene. Cancer Gene Ther 2000;7:284–291.
121 Zhang T, Lemoi BA, Sentman CL: Chimeric NK-receptor-bearing T cells mediate antitumor immunotherapy. Blood 2005;106:1544–1551.
122 Weijtens ME, Willemsen RA, Valerio D, Stam K, et al: Single chain Ig/gamma gene-redirected human T lymphocytes produce cytokines, specifically lyse tumor cells, and recycle lytic capacity. J Immunol 1996;157:836–843.
123 Rossig C, Bollard CM, Nuchtern JG, Merchant DA, et al: Targeting of G_{D2}-positive tumor cells by human T lymphocytes engineered to express chimeric T-cell receptor genes. Int J Cancer 2001;94:228–236.
124 Abken H, Hombach A, Heuser C, Reinhold U: A novel strategy in the elimination of disseminated melanoma cells: chimeric receptors endow T cells with tumor specificity. Recent Results Cancer Res 2001;158:249–264.
125 Gonzalez S, Naranjo A, Serrano LM, Chang WC, et al: Genetic engineering of cytolytic T lymphocytes for adoptive T-cell therapy of neuroblastoma. J Gene Med 2004;6:704–711.
126 Morgenroth A, Cartellieri M, Schmitz M, Gunes S, et al: Targeting of tumor cells expressing the prostate stem cell antigen (PSCA) using genetically engineered T-cells. Prostate 2007;67:1121–1131.
127 Hwu P, Yang JC, Cowherd R, Treisman J, et al: In vivo antitumor activity of T cells redirected with chimeric antibody/T-cell receptor genes. Cancer Res 1995;55:3369–3373.
128 Kershaw MH, Westwood JA, Hwu P: Dual-specific T cells combine proliferation and antitumor activity. Nat Biotechnol 2002;20:1221–1227.
129 Hekele A, Dall P, Moritz D, Wels W, et al: Growth retardation of tumors by adoptive transfer of cytotoxic T lymphocytes reprogrammed by CD44v6-specific scFv:ζ-chimera. Int J Cancer 1996;68:232–238.
130 Dall P, Herrmann I, Durst B, Stoff-Khalili MA, et al: In vivo cervical cancer growth inhibition by genetically engineered cytotoxic T cells. Cancer Immunol Immunother 2005;54:51–60.
131 Pameijer CR, Navanjo A, Meechoovet B, Wagner JR, et al: Conversion of a tumor-binding peptide identified by phage display to a functional chimeric T cell antigen receptor. Cancer Gene Ther 2007;14:91–97.
132 Gilham DE, O'Neil A, Hughes C, Guest RD, et al: Primary polyclonal human T lymphocytes targeted to carcino-embryonic antigens and neural cell adhesion molecule tumor antigens by CD3ζ-based chimeric immune receptors. J Immunother 2002;25:139–151.
133 Niederman TM, Ghogawala Z, Carter BS, Tompkins HS, et al: Antitumor activity of cytotoxic T lymphocytes engineered to target vascular endothelial growth factor receptors. Proc Natl Acad Sci U S A 2002;99:7009–7014.
134 Gattenlohner S, Marx A, Markfort B, Pscherer S, et al: Rhabdomyosarcoma lysis by T cells expressing a human autoantibody-based chimeric receptor targeting the fetal acetylcholine receptor. Cancer Res 2006;66:24–28.
135 Brentjens RJ, Santos E, Nikhamin Y, Yeh R, et al: Genetically targeted T cells eradicate systemic acute lymphoblastic leukemia xenografts. Clin Cancer Res 2007;13:5426–5435.
136 Loskog A, Giandomenico V, Rossig C, Pule M, et al: Addition of the CD28 signaling domain to chimeric T-cell receptors enhances chimeric T-cell resistance to T regulatory cells. Leukemia 2006;20:1819–1828.
137 Vera J, Savoldo B, Vigouroux S, Biagi E, et al: T lymphocytes redirected against the κ light chain of human immunoglobulin efficiently kill mature B lymphocyte-derived malignant cells. Blood 2006;108:3890–3897.
138 Di Stasi A, De Angelis B, Rooney CM, Zhang L, et al: T lymphocytes coexpressing CCR4 and a chimeric antigen receptor targeting CD30 have improved homing and antitumor activity in a Hodgkin tumor model. Blood 2009;113:6392–6402.
139 Dutour A, Marin V, Pizzitola I, Valsesia-Wittmann S, et al: In vitro and in vivo antitumor effect of anti-CD33 chimeric receptor-expressing EBV-CTL against CD33 acute myeloid leukemia. Adv Hematol 2012;2012:683065.

140 Thokala R, Olivares S, Singh H, Champlin RE, Cooper LJN: Targeting leukemia by CD123 specific chimeric antigen receptor. Biol Blood Marrow Transplant 2012;18:S291.
141 Pinthus JH, Waks T, Malina V, Kaufman-Francis K, et al: Adoptive immunotherapy of prostate cancer bone lesions using redirected effector lymphocytes. J Clin Invest 2004;114:1774–1781.
142 Davies DM, Foster J, van der Stegen SJC, Parente-Pereira AC, et al. Flexible targeting of ErbB dimers that drive tumorigenesis by using genetically engineered T cells. Mol Med 2012;18:565–576.
143 Haynes NM, Trapani JA, Teng MW, Jackson JT, et al: Rejection of syngeneic colon carcinoma by CTLs expressing single-chain antibody receptors codelivering CD28 costimulation. J Immunol 2002; 169:5780–5786.
144 Maher J, Brentjens RJ, Gunset G, Riviere I, et al. Human T-lymphocyte cytotoxicity and proliferation directed by a single chimeric TCRζ/CD28 receptor. Nat Biotechnol 2002;20:70–75.
145 Zhang T, Wu MR, Sentman CL: An NKp30-based chimeric antigen receptor promotes T cell effector functions and antitumor efficacy in vivo. J Immunol 2012;189:2290–2299.
146 Kong S, Sengupta S, Tyler B, Bais AJ, et al: Suppression of human glioma xenografts with second-generation IL13R-specific chimeric antigen receptor-modified T cells. Clin Cancer Res 2012;18: 5949–5960.
147 Chekmasova AA, Rao TD, Nikhamin Y, Park KJ, et al: Successful eradication of established peritoneal ovarian tumors in SCID-beige mice following adoptive transfer of T cells genetically targeted to the MUC16 antigen. Clin Cancer Res 2010;16: 3594–3606.
148 Kailayangiri S, Altvater B, Meltzer J, Pscherer S, et al: The ganglioside antigen G_{D2} is surface-expressed in Ewing sarcoma and allows for MHC-independent immune targeting. Br J Cancer 2012; 106:1123–1133.
149 Westwood JA, Smyth MJ, Teng MW, Moeller M, et al: Adoptive transfer of T cells modified with a humanized chimeric receptor gene inhibits growth of Lewis-Y-expressing tumors in mice. Proc Natl Acad Sci U S A 2005;102:19051–19056.
150 Willemsen RA, Ronteltap C, Chames P, Debets R, et al: T cell retargeting with MHC class I-restricted antibodies: the CD28 costimulatory domain enhances antigen-specific cytotoxicity and cytokine production. J Immunol 2005;174:7853–7858.
151 Cheung NK, Guo HF, Modak S, Cheung IY: Anti-idiotypic antibody facilitates scFv chimeric immune receptor gene transduction and clonal expansion of human lymphocytes for tumor therapy. Hybrid Hybridomics 2003;22:209–218.
152 Kochenderfer JN, Feldman SA, Zhao YB, Xu H, et al: Construction and preclinical evaluation of an anti-CD19 chimeric antigen receptor. J Immunother 2009;32:689–702.
153 Marin V, Pizzitola I, Agostoni V, Attianese GMPG, et al: Cytokine-induced killer cells for cell therapy of acute myeloid leukemia: improvement of their immune activity by expression of CD33-specific chimeric receptors. Haematologica 2010;95:2144–2152.
154 Morgan RA, Johnson LA, Davis JL, Zheng Z, et al: Recognition of glioma stem cells by genetically modified T cells targeting EGFRvIII and development of adoptive cell therapy for glioma. Hum Gene Ther 2012;23:1043–1053.
155 Carpenito C, Milone MC, Hassan R, Simonet JC, et al: Control of large, established tumor xenografts with genetically retargeted human T cells containing CD28 and CD137 domains. Proc Natl Acad Sci U S A 2009;106:3360–3365.
156 Zhong XS, Matsushita M, Plotkin J, Riviere I, et al: Chimeric antigen receptors combining 4-1BB and CD28 signaling domains augment PI3kinase/AKT/Bcl-XL activation and CD8+ T cell-mediated tumor eradication. Mol Ther 2010;18:413–420.
157 Wilkie S, Picco G, Foster J, Davies DM, et al: Retargeting of human T cells to tumor-associated MUC1: the evolution of a chimeric antigen receptor. J Immunol 2008;180:4901–4909.
158 Geldres C, Savoldo B, Hoyos V, Caruana I, et al: T lymphocytes redirected against the chondroitin sulfate proteoglycan-4 control the growth of multiple solid tumors both in vitro and in vivo. Clin Cancer Res 2014;20:962–971.

Ulf Petrausch, MD
Department of Oncology, University Hospital Zürich
Rämistrasse 100
CH–8091 Zurich (Switzerland)
E-Mail ulf.petrausch@usz.ch

Author Index

Subject Index